H. S. Carhart,

Feb. 22, 1881.

Evanston, Ill.

GENERAL OFFICE OF THE WESTERN UNION TELEGRAPH COMPANY, CORNER BROADWAY AND LIBERTY STREET, NEW YORK.

MODERN PRACTICE

OF THE

ELECTRIC TELEGRAPH.

A HANDBOOK

FOR

ELECTRICIANS AND OPERATORS.

BY FRANK L. POPE.

SEVENTH EDITION.
REVISED AND ENLARGED.

New York:
D. VAN NOSTRAND, Publisher,
23 MURRAY STREET & 27 WARREN STREET.

1872.

PREFACE TO THE FOURTH EDITION.

During the quarter of a century which has elapsed since the introduction of the Electric Telegraph in the United States, those engaged in its service have been almost entirely dependent upon verbal instruction, and long practical experience, for a thorough technical knowledge of their profession. The works accessible to the American telegrapher have been of a popular, rather than of a strictly scientific character, or else of so elementary a nature as to be of little service except to the most inexperienced students. It is true that a number of excellent foreign works have appeared within a few years ; yet the difficulty and expense of obtaining them, as well as their want of applicability to the American telegraphic system, has prevented their general circulation among the class for which this work is more especially designed.

The unexpectedly favorable reception which has been accorded to the first three editions of this work, has led the author to believe that it has, to some extent, supplied the acknowledged deficiency which had previously existed in this branch of literature. The present edition has been carefully revised, as well as enlarged by the addition of much new matter, and is believed to embrace all the recent discoveries and improvements in practical telegraphy, which have successfully passed through the test of actual experience.

The methods of testing telegraph lines and apparatus by actual measurement, which are now universally employed in Europe, and to some extent in this country, have been treated upon to an extent commensurate with the importance of the subject. It is hoped that, with the aid of this work, the student may obtain a complete and satisfactory knowledge of this useful and beautiful system.

The principles laid down for the guidance of the student in the formation of the telegraphic alphabet, and the subsequent progressive exercises intended for practice with the key, differ but slightly from those employed by the author, while teaching a class of students for the American Telegraph Company in 1864. This plan was believed at that time to be original, but as a method of teaching, involving substantially the same principles, was devised and subsequently published by Prof. J. E. Smith, in his *Manual of Telegraphy*, it seems proper to make this explanation of the circumstances.

Among the additional matter in the present edition will be found an entire new chapter upon the Recent Improvements in Telegraphic Practice, as well as a number of articles in the Appendix, on the Equipment of Telegraph Lines, the Working Capacity of Telegraph Lines, and the Electrical Tension of Batteries and Lines, etc., etc.

Most of the illustrations in this volume have been engraved expressly for its pages, from original drawings by the author.

In conclusion, the author desires to express his acknowledgments to his friend David Brooks, for much valuable aid in the preparation of this work, especially of the present edition ; and he would likewise take occasion to thank M. G. Farmer, for information which has been kindly supplied by him. Much useful material has also been obtained from Sabine's *Electric Telegraph*, Culley's *Hand-Book of the Electric Telegraph*, Clark's *Electrical Measurement*, Varley's *Report on the Condition of the Western Union Lines*, and the columns of *The Telegrapher*.

ELIZABETH, N. J., *January*, 1871.

CONTENTS.

CHAPTER I

ORIGIN OF THE ELECTRIC CURRENT.—GALVANIC BATTERIES.

CHAPTER II.

ELECTRO-MAGNETISM.

CHAPTER III.

TELEGRAPHIC CIRCUITS.

CHAPTER IV.

THE MORSE, OR AMERICAN TELEGRAPHIC SYSTEM.

CHAPTER V.

INSULATION.

CHAPTER VI.

TESTING TELEGRAPH LINES.

CHAPTER VII.

NOTES ON TELEGRAPHIC CONSTRUCTION.

CHAPTER VIII.

HINTS TO LEARNERS.

CHAPTER IX.

RECENT IMPROVEMENTS IN TELEGRAPHIC PRACTICE.

CHAPTER X.

APPENDIX AND NOTES.

MODERN PRACTICE

OF THE

ELECTRIC TELEGRAPH.

CHAPTER I.

ORIGIN OF THE ELECTRIC CURRENT.—GALVANIC BATTERIES.

1. Simple Galvanic Circuit.—If two plates of different metals, such as copper and zinc for example, are immersed in a vessel of water to which a small portion of sulphuric acid has been added, and the upper ends of the two plates are brought in contact, or connected together with a metallic wire as in fig. 1, a continuous

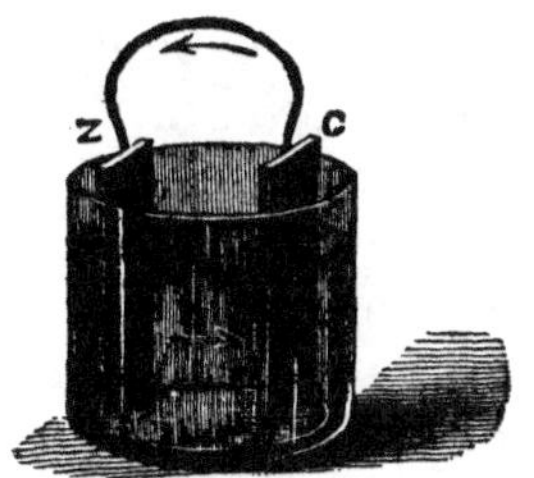

Fig. 1.

current of electricity will pass from the copper to the zinc through the connecting wire, and from the zinc to the copper through the liquid, as indicated by the arrows in the figure. If the metallic communication be interrupted, or the *circuit*, as it is termed, broken, the current at once ceases, but is instantly renewed whenever the connection is again formed. Electricity produced by this means is usually termed Galvanic or Voltaic electricity, from the names of its discoverers, and is the effect of chemical action by the acidulated water upon the zinc.

2. The plate (usually of zinc), upon the surface of which the electricity is generated by chemical action, is called the *negative pole*, and the opposite plate, generally of copper, platina or carbon, is called the *positive pole*. They are also frequently designated by the signs — (minus) and + (plus).

3. If both metals in this arrangement were equally acted upon by the solution, no electricity would be produced, as this effect arises in all cases from the difference in the chemical action upon the two plates. For this reason the positive plate is made of some metal or other substance upon which the liquid has little or no effect.

4. The apparatus for producing voltaic electricity, which has been described in its simplest form, is called a *battery*. As electricity is produced under any circumstances in which the above conditions have been complied with, there are various methods of constructing a battery. The forms used in the practical operation of the telegraph will hereafter be described in detail.

5. CONDUCTORS AND NON-CONDUCTORS.—Some substances, such as metals, possess the property of allowing electricity to diffuse itself freely throughout their whole substance, and are therefore termed *conductors*. Others, such as glass, hard rubber, and dry wood, offer great resistance or opposition to this diffusion, and are called *non-conductors* or *insulators*.

6. This division however is relative and not absolute. Few if any bodies are perfect insulators, and even metals, the most perfect of all conductors, offer *some* resistance to the passage of electricity, or in other words insulate slightly. A good insulator, therefore, is simply a bad conductor, and *vice versa*.

7. In the following list each substance named conducts better than that which precedes it, the first being the best insulator and the last the best conductor:

1. Dry Air,	5. India Rubber,	9. Silk,
2. Paraffine,	6. Gutta Percha,	10. Dry Paper,
3. Hard Rubber,	7. Sulphur,	11. Porcelain,
4. Shellac,	8. Glass,	12. Dry Wood,

13. Dry Ice,
14. Water,
15. Saline Solutions,
16. Acids,
17. Charcoal or Coke,
18. Mercury,
19. Lead,
20. Tin,
21. Iron,
22. Platinum,
23. Zinc,
24. Gold,
25. Copper,
26. Silver.

8. ELECTRICAL TENSION.—If two or more simple batteries, or *elements* as they are called, are connected together in such a manner that the positive plate of the first is united by a metallic conductor with the negative plate of the second, and so on, as shown in fig. 2, the

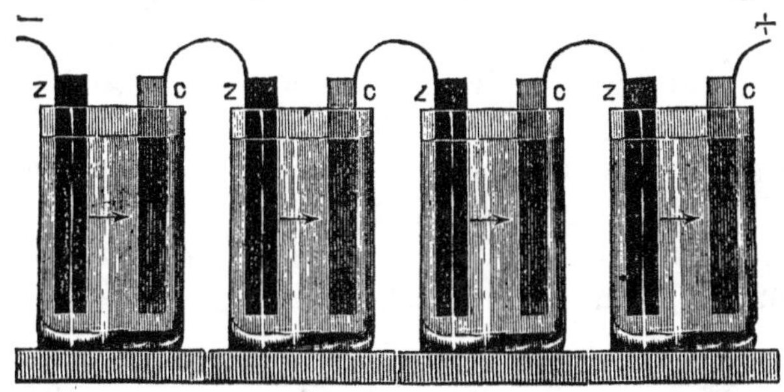

FIG. 2.

electrical *tension*, or power of overcoming resistance, is increased in direct proportion to the number of elements. Four elements will therefore possess four times the tension of one element, and the current generated by their combined action will be capable of overcoming four times the resistance of that from a single element.

9. ELECTRICAL QUANTITY.—It is important, however, to observe, that although the *tension* increases with each element added to the series, no greater *quantity* is produced by a great number of elements than by a single one—the action in each cell serving only, as it were, to *urge forward* a quantity equal to that arising from chemical decomposition in the first cell. If, on the contrary, we connect together the four zincs and the four coppers, forming in effect a single element, with plates equivalent to four times the original surface, there will be four times the original quantity of electricity generated; but its tension, or power of overcoming resistance, will be no greater than that of a single pair of plates. This distinction is of great importance,

and should be thoroughly understood and carefully remembered.

10. In the simple form of battery previously described (8), if the poles are united by a conductor for a considerable length of time, bubbles of hydrogen, arising from the decomposition of the water, cover the positive plate, and in a great measure prevent the liquid from coming in contact with it, and the surface of the plate also becomes coated with a deposit of zinc, tending to convert the battery into one in which both plates are of zinc, and thus its electro-motive force is weakened and finally destroyed. In order to render the battery *constant* in its action, it is necessary to prevent these effects by surrounding the negative plate with a solution of a salt of the metal itself. This principle is employed in the arrangement about to be described.

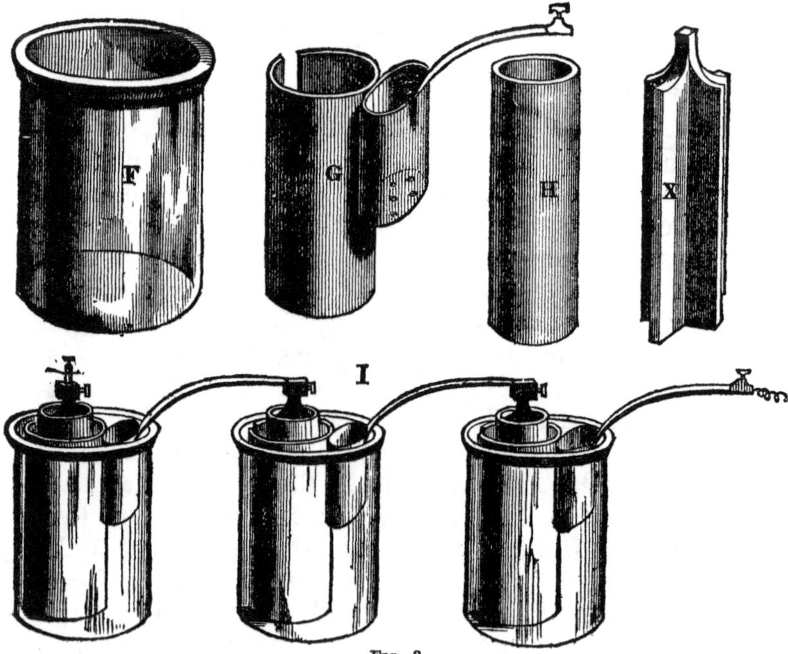

Fig. 3.

11. The Daniell Battery.—This combination consists of a jar of glass or earthenware, F (fig. 3), about six

inches in diameter and eight or nine inches high. A plate of copper, G, is bent into a cylindrical form, so as to fit within it, and is provided with a perforated chamber, to contain a supply of sulphate of copper in crystals, and a strap of the same metal with a clamp for connecting it to the zinc of the next element. H is a *porous cup*, as it is technically termed, made of unglazed earthenware, six or seven inches high and two inches in diameter, within which is placed the zinc, X. This is usually of the shape shown in the figure, which is called the "star zinc," but it is often made in the form of a hollow cylinder, the latter giving greater power, but being somewhat more difficult to clean.

The outer cell is filled with a saturated solution of sulphate of copper (blue vitriol), and the porous cell with a solution of sulphate of zinc. A series of three elements connected together, as usually employed on American lines for a *local battery*, is shown at I.

12. Effect of Continued Action.—By continued action sulphate of zinc is formed in the porous cup, and the sulphate of copper in the outer cell consumed, the zinc being constantly dissolved away while the copper plate is at the same time increased. When all the sulphate of copper has been decomposed, and the water in the zinc compartment saturated with sulphate of zinc, the action of the battery ceases. Some of the sulphate of zinc in this case usually passes into the copper cell, and appears upon the copper plate in the form of a black powder ; it is therefore necessary to maintain a constant supply of pulverized vitriol in the perforated chamber attached to the copper cylinder.

13. When the solution in the porous cup becomes saturated with sulphate of zinc it crystallizes upon the zinc plate, interfering with the action of the battery. Part of this solution should therefore be removed occasionally and replaced with water.

In setting up the battery pure water may be used in the porous cell, and the battery allowed to stand a few hours with a closed circuit, when it will be found ready for use. The addition of a little sulphate of zinc will greatly hasten its action.

14. THE DEPOSIT OF COPPER UPON THE POROUS CUP.—This cannot be entirely prevented, but may be greatly lessened by suspending the zinc so that it will not touch the porous cup below the surface of the liquid, and by saturating the bottom of the cell to the height of half an inch with melted paraffine, or even tallow.

15. When constructed as above described and used in a local circuit, the Daniell battery will continue in action about ten or fifteen days without attention, the time depending upon the size of the wire in the magnet and the amount of daily service. The sulphate of copper solution should be kept of good strength, otherwise the upper portion becomes weak and an extra current is set up within the battery, which tends to eat away and destroy the copper plate without any useful effect.

16. RENEWAL OF THE BATTERY.—In renewing this battery the zincs should be scraped and well cleaned with a stiff brush, the porous cups thoroughly washed, and the old solution contained in them thrown out, with the exception of about one third of the clear portion, which should be returned, otherwise the battery will require some hours to recover its full strength. The copper deposit upon the zincs is valuable, and should be preserved.

Every two or three months the coppers ought to be taken out and the deposit upon their surface removed, which may be done two or three times. When they become too much encrusted to afford room for the porous cups they must be replaced by new ones.

Porous cups ought to be renewed whenever they become too much encrusted with copper. If cracked they should be changed at once, otherwise a great waste of material will ensue.

17. The crystals which form around the edge of the outer jar require to be occasionally wiped off with a damp cloth, or they will eventually run down the outside and form a connection between the jars, giving rise

to a great consumption of material without corresponding benefit.

18. In order that the current may act with its full force, it is necessary to keep the clamps and connections of the battery clean and bright, and free from rust or dirt. As chemical action is promoted by heat, the battery will act more vigorously if kept in a warm place.

19. Application of the Daniell Battery to Main Circuits.—This battery is sometimes used for main circuits, but in that case it is preferable to arrange it differently by placing the zincs outside and the copper within the porous cell, as in fig. 4, in which Z shows the

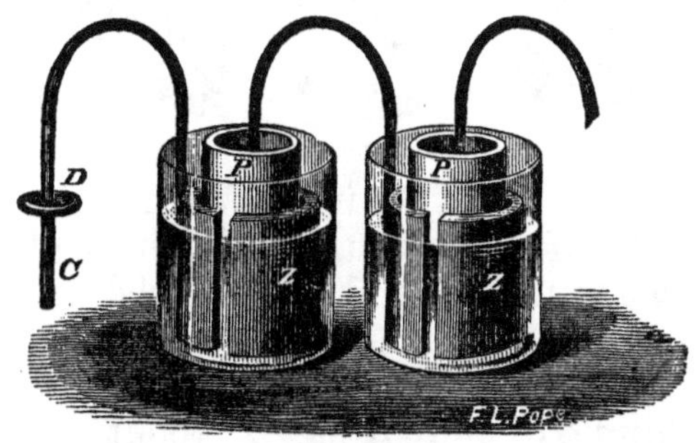

Fig. 4.

zinc and P the porous cell. The copper, C, is provided with a perforated shelf, D, upon which the vitriol is placed.

Other forms have been devised which dispense entirely with the porous cup, the two solutions being separated by the difference in their respective specific gravities. Some of these bid fair to come into extensive use.

20. The Grove Battery.—The most intense and powerful voltaic combination that has yet been discovered is that of Grove. For many years it was exclusively used for telegraphic purposes in this country, and is still employed in that capacity to a considerable extent. Its component parts are shown in fig. 5, in which A represents a glass jar or tumbler, about 3

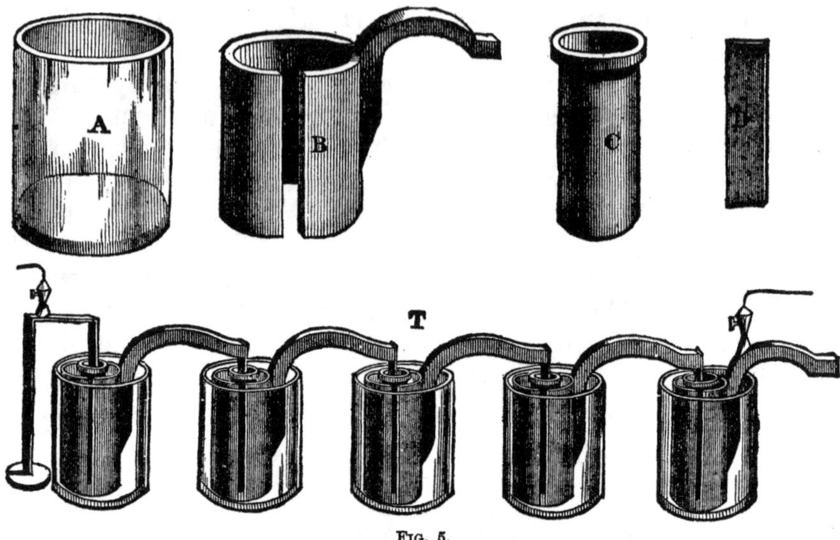

Fig. 5.

inches in diameter and 4½ inches high. A thick cylinder of zinc, B, of a size nearly sufficient to fill the tumbler, is placed within it, and is furnished with a projecting arm, to which is attached the positive plate of the next element. The porous cup, C, is placed within the zinc. A thin strip of platina, D, about 2½ inches long and half an inch in width, is soldered to the end of the zinc arm projecting from the adjacent cell, and reaches nearly to the bottom of the porous cup.

21. Setting up a Grove Battery.—It is necessary that the zinc should first be thoroughly *amalgamated*. The ordinary zinc of commerce contains particles of lead, iron, and other impurities, which, when the plate is immersed in dilute acid, form as it were small batteries upon the surface, which eat away numerous cavities in the zinc without producing any useful effect. This is termed *local action*, and may be, in a great measure, prevented by the above process of amalgamation, which is usually performed by immersing the zincs in a vessel containing dilute muriatic or sulphuric acid, and then plunging them in a bath of metallic mercury. After remaining in this for a minute or two they are taken

out and placed in a vat of clean water, where the superfluous mercury is allowed to drain off. The mercury dissolves a little of the zinc, which flows over and covers the impurities, and prevents the acid solution from coming in contact with them.

22. In putting the Grove battery together, first place the glass tumblers in position and fill them about half full of a solution composed of one part of sulphuric acid and twenty to thirty parts water, by measure, thoroughly mixed. Then place the amalgamated zincs in the tumblers, with the arms turned at right angles to the line of cells. Fill the porous cups nearly full of strong nitric acid and place them within the zincs, then turn the zincs around so as to immerse the platina strips in the nitric acid of the adjoining cell, throughout the whole series, as shown at T, in fig. 5.

23. The strength of the dilute sulphuric acid solution in this battery should be varied in proportion to the number of wires worked from it. The less the number of the latter the weaker the solution may be made.

24. When in continuous service a Grove battery ought to be taken apart every night, and the nitric acid from the porous cups emptied into a vessel and kept closed until morning. The zincs should be removed and placed inverted in a trough of water, acidulated with sulphuric acid, and in the morning rubbed with a brush, and the mercury diffused evenly over their surfaces. To every ten parts of the nitric acid taken from the battery add one part of fresh acid every morning. By this means a steady and uniform current will be maintained when the battery is in action. The dilute sulphuric acid requires renewal about twice a week. In handling this battery great care is required not to injure the connection between the zinc and the platina. A set of Grove zincs, in continuous service, will require renewal about once in three months.

25. The Carbon Battery.—This is a modification of the Grove battery, and is sometimes called the Electropoion battery. It is extensively employed on the Ame-

rican lines for main circuits. In its general construction and arrangement it differs but little from the battery last described. The different parts of which it is composed are shown in fig. 6, consisting of a glass tumbler, zinc and porous cup. In place of the platina of the Grove battery, a plate of carbon or coke is employed for the positive element, as shown in the figure.

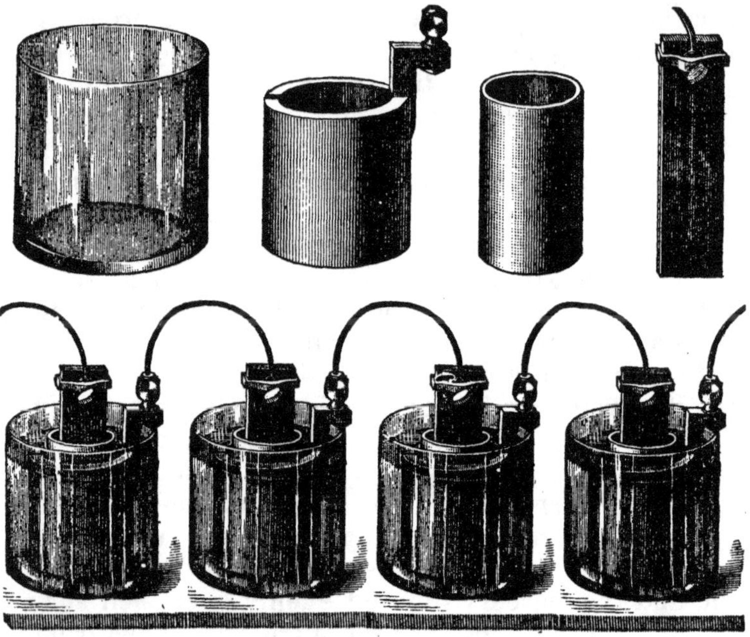

Fig. 6.

A clamp is arranged so as to press a platina button firmly against the carbon, this button being permanently attached to a wire leading to a binding screw on the zinc arm of the next element. The parts are usually made of about the same size as in the Grove battery.

The carbon connection is sometimes made by means of a platinized copper wire inserted into its upper end, and surrounded with lead, to prevent the action of the acids upon the copper.

26. In setting up this battery the different parts

should be put together in the position they are to occupy, as shown in fig. 6, and care taken that all the connections are firmly screwed up. The zincs must be thoroughly amalgamated, and the dilute sulphuric acid solution mixed as directed for the Grove battery. A sufficient quantity of this solution is poured into the tumblers to cover the cylindrical portion of the zincs. The porous cups are then filled with a solution of bichromate of potash,* care being taken not to pour it upon the connections or clamps.

27. When the battery is in service, one third of the bi-chromate solution in the porous cups should be removed every morning by means of a large rubber syringe, and replaced with fresh. A new set of zincs will require to be amalgamated a second time after having been in use three or four days; after which once in two to four weeks will be often enough—depending somewhat upon the amount of work required from the battery. The battery ought to be taken apart every two weeks, the zincs brushed, the dilute sulphuric acid solution renewed, and the carbons thoroughly soaked in clean water. It is better, if possible, to have a spare set of cells complete, so that one may be renewed while the other is in use.

28. Power of the Carbon Battery.—This is quite equal to that of the Grove, as far as the intensity of its action is concerned. The latter however will work nearly twice as many wires at the same time as the former. The expense of the carbon battery for materials and attendance is less than one third that of the Grove. A set of zincs, if properly cared for, will last from fourteen to sixteen months on an ordinary telegraph line. It is a good plan to coat the zincs with asphaltum varnish at the junction of the projecting arm, as these are frequently eaten off while the rest of the zinc remains in good condition.

* This solution is made as follows: Mix one gallon of sulphuric acid and three gallons of water. Then, in a separate vessel, dissolve five lbs. bi-chromate of potash in two gallons of boiling water and add to the above, mixing the whole thoroughly together. The proportion of bi-chromate is sometimes made one fifth greater than the amount given.

29. Insulation of Batteries.—The cells of a battery should always be thoroughly insulated from each other. This is especially important in the case of the Grove battery. A convenient and effective mode of insulation is shown in fig. 7, in which the battery tumblers, AA, are set upon hollow cylinders of wood, BB, saturated with asphaltum or paraffine, and insulated from the upright wooden pins, DD, by the glass sockets, CC. The pins are inserted into a horizontal scantling, E, which forms the top of the battery stand.

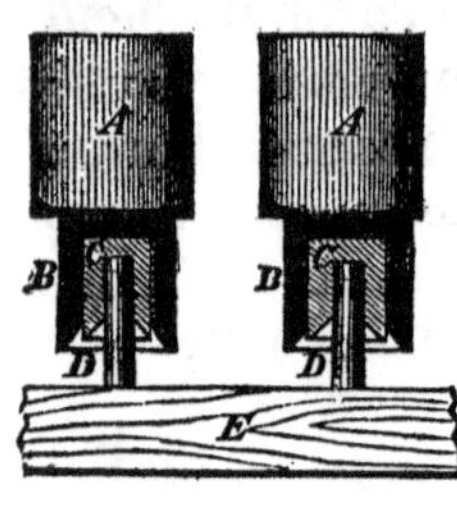

Fig. 7.

Battery jars, of different sizes, are now made at the Brooks Paraffine Insulator Works, in Philadelphia, which are composed of stone-ware, thoroughly saturated with paraffine, so that moisture will not penetrate them, nor remain upon their surface. When these jars are employed, no special insulation is required.

CHAPTER II.

ELECTRO-MAGNETISM.

30. WHENEVER the poles of a battery are connected by a conductor, or series of conductors, so as to form a *circuit*, a *current* of electricity is assumed to flow from the negative to the positive pole, through the battery itself, and from the positive to the negative pole through the conductor.

31. If the conducting wire is covered with an insulator (5), such as silk or cotton, so as to compel the current to traverse its entire length, and is wound into a spiral or coil, surrounding a magnetic needle, the needle will be deflected from its natural position, and will tend to take up a position at right angles to the direction of the current. If the current be passed in the opposite direction through the wire, the deflection of the needle will also take place in the opposite direction. The *Galvanometer*, an extremely useful instrument for the purpose of indicating the presence, direction, and strength of a voltaic current, is constructed upon this principle.

32. If the conducting wire, covered as above, be wound upon a bar of soft iron, the iron becomes *magnetic* as long as the current continues to flow, and possesses the property of attracting other pieces of iron in its vicinity. This arrangement is called an *electro-magnet.* (34.)

33. If the iron is very soft and pure it loses its magnetism instantly upon the cessation of the current, but if impure, or if hardened by hammering or turning, it retains a certain amount of *residuary magnetism*, especially after it has been acted upon by a powerful current. It is, therefore, necessary that the iron *cores*, as they are termed, of electro-magnets, should be annealed with great care.

34. Electro-Magnets are generally made in a U form, two bobbins or spools, *a a* (fig. 8), being filled with covered copper wire, and the soft iron cores, *c c*, passing through them, fixed upon a connecting bar, *b*, also of soft iron, as shown in the figure. The two spools, *a* and *a*, are virtually continuations of one spool, the direction being apparently reversed by the bend of the U. The ends of the cores, *c c*, opposite to the connecting bar, are called the *poles* of the magnet, the magnetic force being accumulated at these points. The bar of soft iron, *d*, upon which the magnet exerts its force, is called the *armature*.

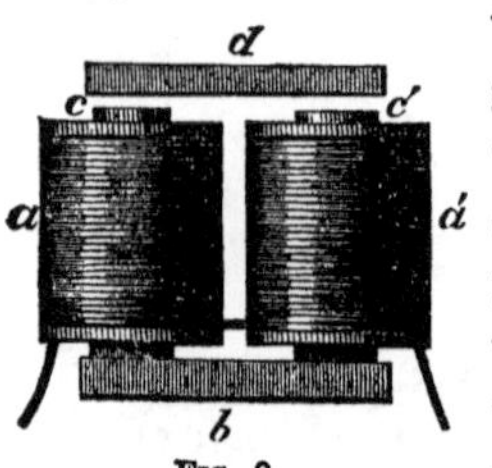

Fig. 8.

35. In electro-magnets and galvanometers the magnetic effect of the current is multiplied by the number of convolutions of the wire in the coil, but it is diminished in proportion to the distance of the wire from the core, each layer of wire acting with less power than the one beneath it.

36. Every addition to the length of the conducting wire enfeebles the current, because of the increased resistance (5, 6,) it offers to its passage. In a very long circuit, such as a telegraph line, the action of the current will necessarily be feeble, and the coil is, therefore, made of *fine* wire, which occupies little space, and allows many layers to be wound on without too greatly increasing the distance from the cores, while its resistance is too small in proportion to the rest of the circuit to reduce the strength of the current materially.

37. When, however, the circuit is very short, coarser wire is employed in the coil. A fine wire would add to the resistance of the circuit more than would be made up by the effect of an increased number of turns, for even a very few layers would double the resistance of the circuit.

The former is frequently called an *intensity*, and the latter a *quantity magnet.*

38. Iron does not acquire its full magnetism instantaneously, and the act of demagnetization also requires time, but is effected more rapidly than magnetization. The greater the tension of the battery the more rapidly the iron acquires its magnetism; therefore, if very rapid action is required, even on a short circuit, a number of cells of battery must be used.

It has also been ascertained by experiment that an electro-magnet with short cores, will acquire and lose its magnetism with much greater rapidity than one with long cores, but in other respects similar.

CHAPTER III.

TELEGRAPHIC CIRCUITS.

39. A Telegraphic Circuit consists of one or more batteries, the line wire, the instruments and the earth. When the circuit is very short a return wire is frequently used instead of the earth.

40. Owing to the immense rapidity with which the electric force is propagated throughout a circuit, any effect which can be produced at hand can be produced in any other part of a circuit, however distant, at the same instant of time, subject to a diminution of force, arising from causes which diminish the quantity of electricity, or the force of the current before its arrival at the distant end, thus weakening its effect. The principal causes of this diminution are the resistance of the circuit and defective insulation, in consequence of which a portion of the current escapes from the line to the earth, and returns without traversing the distant portion of the circuit.

41. The *effective force* of the current leaving the battery depends upon two things—the *tension* of the battery, which sets the current in circulation, and the *resistance* the current encounters in traversing the circuit.

42. Resistance of the Circuit.—This depends upon the length and size of the conductor, and the material of which it is composed. In an ordinary telegraphic line wire the resistance is in direct proportion to its length, and also in inverse proportion to its weight per mile. Thus, 150 miles of No. 8 wire will conduct as well as 100 miles of No. 10 wire, and as great an effect can be produced at its remote end with a battery of equal tension. There is, therefore, a great advantage

in using the larger sizes of wire in the construction of lines intended to be worked in long circuits.

43. Electrical Measurement.—In order to institute a comparison between the resistances of different circuits, etc., a standard has been fixed upon by the British Association, called the *Ohm*, which is equivalent to about $\frac{1}{16}$ of a mile of galvanized No. 9 iron wire, such as is usually employed in the construction of telegraph lines. This standard unit of resistance is now made use of by the English electricians.

44. Resistance Coils.—As no battery is constant in its power, and no magnet uniform in its strength, neither of these can be made use of as an accurate basis of comparison. *Resistance coils*, composed of wire of certain alloys of metals, carefully prepared, have been found not to vary $\frac{1}{1,000,000}$ in eight years. The only variation is that due to difference in temperature, which may be readily calculated and allowed for when necessary.

It will, therefore, be understood that the ohm is a unit of resistance in the same manner that an inch is a unit of length, or a pound a unit of weight.

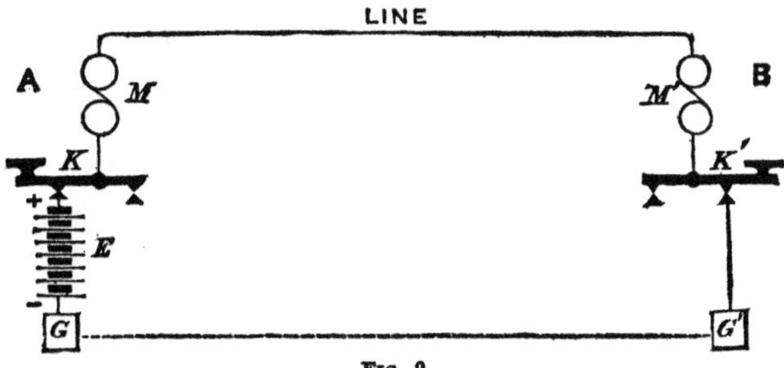

Fig. 9.

45. A Telegraphic Circuit, in its simplest form, is shown in fig. 9. A and B represent two stations. The circuit may be traced as follows: From the + pole of the battery E to the key K (52) and electro-magnet M, thence through the line L to the other station, electro-magnet M′ and key K′ to the earth at G′, and thence through the earth, as represented by the dotted line, to the – pole of the battery E. A continuous current will

therefore flow through the circuit as long as it remains uninterrupted, and the armatures of the electro-magnets M and M′ will be attracted by the cores, but if the circuit be broken by means of one of the keys, K or K′, both electro-magnets will be demagnetized. Thus, the breaking of the circuit at either station affects the electro-magnets of both, as they are in the same circuit.

46. THE EARTH CIRCUIT.—In thus using the earth as part of the circuit, it is found that it offers, practically, no resistance to the passage of the current. Although comparatively a poor conductor, it is an infinitely large one in proportion to the wire, and, therefore, its resistance is not appreciable (42).

47. ARRANGEMENT OF BATTERIES.—In practice it is usual to divide the battery E into two parts, placing half at each end of the line, for reasons which will hereafter appear. It is important, however, when this is done, that the positive pole of one battery should be connected with the negative pole of the other, otherwise they would neutralize each other, and no effect would be obtained. In such a case the batteries are said to be *reversed.*

48. INTERMEDIATE STATIONS.—It is evident that intermediate stations may be introduced at any point upon the line shown in the above figure, each being provided with an electro-magnet and key, forming part of the circuit, and that the breaking and closing of the circuit at any of these points will affect all the electro-magnets through which it passes, in the same manner and at the same instant of time.

Any desired number of intermediate stations may be placed upon a line until the combined resistance of their electro-magnets reduces the strength of the current below that required for the convenient working of the circuit.

49. THE MORSE SYSTEM.—The principle of the Morse system of telegraphy consists in conveying arbitrary signals by means of the magnetization and demagnetization of an electro-magnet, by the alternate breaking

and closing of a voltaic circuit in the manner above explained. The conventional alphabet used in America for this purpose is given in another part of this work.

50. OTHER TELEGRAPHIC SYSTEMS.—The type printing telegraph, employing the "Combination" instrument of Phelps, is the only system other than the Morse now in use upon the public lines in the United States. The limited extent to which it is employed renders it unnecessary to give a detailed description of its construction and mode of operation in a work of this kind. The electro-chemical telegraph of Bain, and the beautiful type-printing instruments of House and Hughes, were formerly extensively employed in this country. The former has now given place to the Morse, while the two latter have been superseded by the equally rapid and more simple and effective instrument of Phelps.

In addition to these, the magneto-electric dial instrument of Edmands & Hamblet, and the electro-magnetic alphabetical instrument of Chester are finding extensive employment upon private lines, where extreme rapidity of transmission is not required, thus rendering the employment of skilled operators in such cases unnecessary.

CHAPTER IV.

THE MORSE, OR AMERICAN TELEGRAPHIC SYSTEM.

51. THE Morse Telegraphic Apparatus consists of a signal key for breaking and closing the circuit, and an electro-magnet, the armature of which is attached to a lever carrying a steel point or style, which embosses a mark upon a narrow strip of paper, moved uniformly along by clock-work. As long as a current continues to flow through the coils of the electro-magnet the armature is attracted, and a mark is made upon the moving paper. As soon as the circuit is broken the armature ceases to be attracted, and is withdrawn from contact with the paper by means of a spring. The duration of the current, and consequently the length of the mark, depends upon the duration of the contact made by the key.

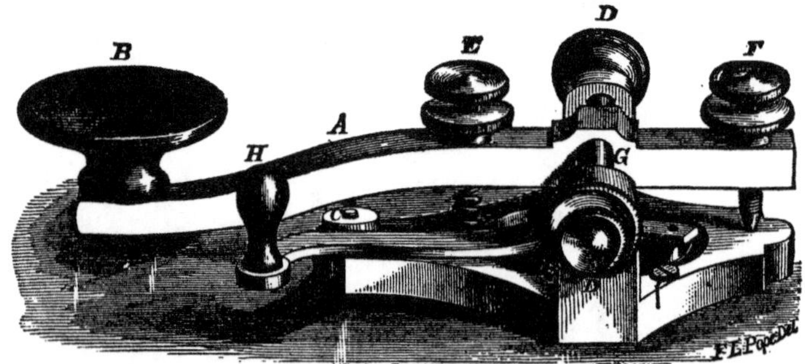

FIG. 10.

52. THE MORSE SIGNAL KEY is shown in fig. 10.* It consists of a brass lever, A, four or five inches in length, which is hung upon a steel arbor, G, between adjustable set screws, D D, in such a manner as to allow it to move freely in a vertical direction. This movement, however, is limited in one direction by the *anvil* C, and in the other by the adjustable set-screw, F.

* The drawings of the signal key, register and relay (figs. 10, 11 and 12), are from instruments manufactured by Bradley.

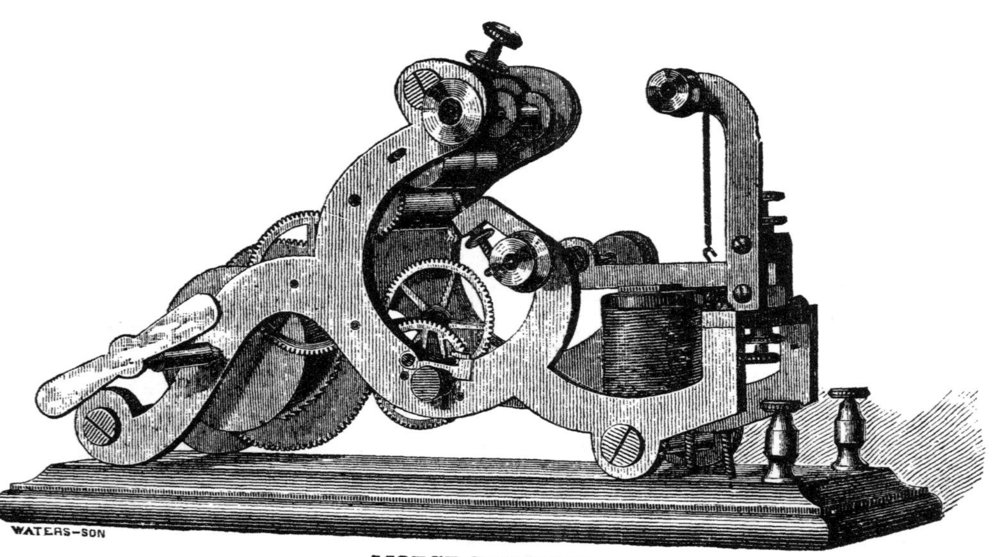

MORSE REGISTER.

Manufactured by L. G. Tillotson & Co., New York.

One wire of the main circuit is connected to the metallic frame of the key, and the other to the anvil, C, which is insulated from the frame. These connections are made by screws passing up through the table from beneath. The lever is provided with a knob of vulcanite, B, by means of which it may be pressed down by the finger of the operator, bringing the lever in contact with the anvil, and thus closing the circuit, precisely as if the wires themselves had been brought together. The points of contact between the lever and the anvil are made of platina, as ordinary metals would be fused by the passage of the electric spark when the circuit is broken. A spring beneath the lever restores it to its original position when the pressure of the operator's finger is withdrawn. When the key is not in use the circuit is completed by bringing the lever of the *circuit closer*, H, into contact with the anvil, C.

FIG. 11.

53. THE MORSE REGISTER.—Fig. 11 represents the recording apparatus, usually termed a *register*, which is made in several different forms, all involving the same principles. M is the electro-magnet, the two ends of the wire forming the coils being carried to the terminal binding screws on the base, one of which is shown at *s*, to which the conducting wires are attached. Above the electro-magnet is seen the armature attached to the lever L, which moves upon an arbor at *d*. The opposite extremity of the lever carries a steel point, *p*.

The strip of paper passes through the guide *g* and between the grooved rollers *r r*, which are moved by a train of wheels driven by a weight attached by a cord to the drum, W.

When the armature is attracted by the magnet the style *p* is brought forcibly in contact with the paper, moving above it upon the grooved roller, and a raised line is embossed upon it corresponding in length to the time the armature remains attracted. A spring adjusted by the nut *n* withdraws the lever when the attraction ceases. The movement of the lever is limited by the adjustable screw, *m*. The screw *c* regulates the pressure of the rollers upon the paper, and the clockwork is started and stopped by the brake *a*. The weight is wound up occasionally, as required, by the operator.

54. The Morse instrument is worked either by the *main line* current or by *relay.* For a distance not exceeding 20 or 30 miles, a register, whose coils are wound with No. 30 copper wire, may be worked by the line current, if the line be well insulated (57).

55. When the insulation is defective, or the circuit so long that its resistance renders the current too weak to work a register direct, as is usually the case with telegraph lines, it becomes necessary to employ a *receiving magnet* or *relay*, which brings a local battery (11) into action at the receiving station, the current of which operates the register.

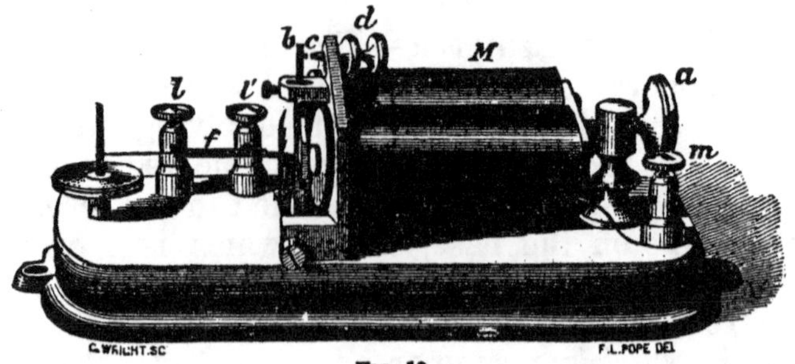

FIG. 12.

56. THE RELAY MAGNET.—The construction of the relay is shown in fig. 12. M is the electro-magnet,

RELAY MAGNET.

Manufactured by Charles Williams, Jr., Boston, Mass.

which is placed in a horizontal position, and is movable by means of the screw *a*. The coils of the magnet are of fine wire, usually from No. 30 to No. 36 in size, of great length and closely wound.* The ends are connected to the line circuit by the binding screws, *m m'*. The armature lever *b* is connected with the binding screw *l* by a wire carried underneath the base of the instrument. A platina point, *c*, on the armature lever, is brought in contact with a similar point on the end of the screw *d* whenever the armature is attracted by the magnet, the screw being in metallic connection with the binding screw *l'*, by means of the frame of the apparatus and a wire beneath the base. One of the screws, *l l'*, is connected to one pole of the local battery (11), and the other to the other pole, embracing the register magnet in its circuit. Therefore, whenever the armature is attracted by the force of the main current acting upon the relay magnet, the circuit of the local battery is completed through the register. As the relay is constructed with great delicacy, a feeble line current is enabled to actuate a register powerfully through the intervention of a local battery.

The movement of the armature is regulated to correspond with the varying strength of the line current by means of the adjustable spiral spring *f*. The magnet may be also set at any required distance from the armature by means of the screw *a*, which is cut with a right and left hand thread, passing through the soft iron bar connecting the two cores, and also through the supporting post in the rear of the coils. The latter slide through openings in the upright metallic plate which supports the adjustable platina pointed screw *d*.

* In the instruments manufactured by Dr. Bradley the helices or coils of the electro-magnets, instead of being composed of silk insulated copper wire, as described in § 34, are made of naked wire, ingeniously wound by accurate machinery in such a manner that the convolutions are separated from each other by a space of 1–600 to 1–800 of an inch, the several layers being insulated from each other by thin paper. It is claimed that, by this method of winding, a coil of a given length and gauge of wire, and, consequently, of a given resistance, can be made of much less diameter than is possible with silk insulated wire, while, at the same time, the number of convolutions will be increased as well as the power of the electro-magnet.

Fig. 13 represents a *Pocket Relay*, as it is usually termed, although it is properly a main line sounder (57).

FIG. 13.

This is provided with a key, as shown in the figure, the whole being conveniently and compactly arranged to fit into an oval case four or five inches long, which may be carried in the pocket. It is an extremely convenient apparatus for line repairers. The cut shows the arrangement manufactured by the Messrs. Chester.

FIG. 14.

57. THE SOUNDER.—In many of the larger telegraph offices the recording apparatus is dispensed with, and the communications read by the sound of the armature lever. In that case the *Sounder* (fig. 14) is employed in the place of the register, the connections of the wires being arranged in precisely the same manner. The Sounder consists simply of the electro-magnet, arma-

ture and lever, fixed upon a base.* The coils are usually wound with No. 23 wire.

Main Line Sounders are used in some offices, which enables the operator to dispense with the local battery. The coils are wound with fine wire, usually No. 30, and are frequently made somewhat larger than those of the relay. A common form of this instrument is known as the "Box Sounder." The lever, striking upon a hollow wooden box containing the magnet, gives a sound that may easily be distinguished by the operator under ordinary circumstances.

FIG. 15.

Fig. 15 (*S. F. Day & Co.*) shows an excellent form of Main Line Sounder. The parts of the instrument are mounted upon a metallic plate, the centre of which is raised slightly above the base, so as to form a bridge, as shown in the cut. The armature lever is of steel, and the whole arrangement is well adapted to increase the sound of the lever as much as possible—a feature of great value in working with weak currents or on badly insulated lines. These instruments are also made in several other forms, and various devices for increasing the sound of the lever are made use of. On many lines they are found to answer as well as the usual arrangement, employing a relay and local battery.

* The instrument shown in the figure is from the manufactory of C. T. & J. N. Chester.

For circuits of moderate length a *Main Line Register* (fig 16), manufactured by Day & Co., has been employed with excellent results.

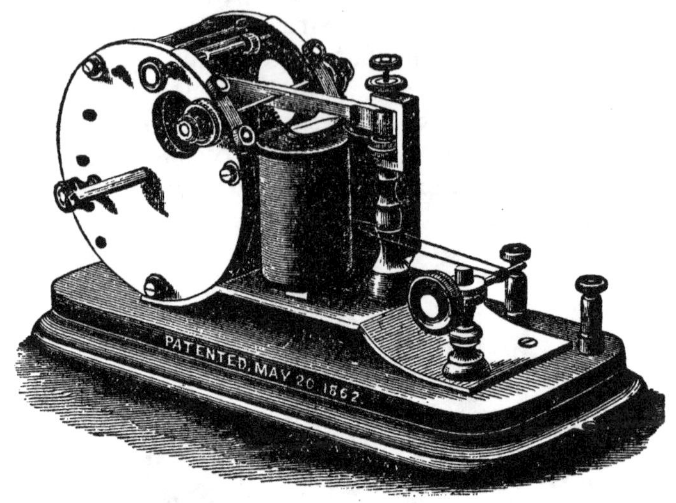

FIG. 16.

58. ARRANGEMENT OF A TERMINAL STATION.—Fig. 17 is a diagram showing the arrangement of wires, batteries,

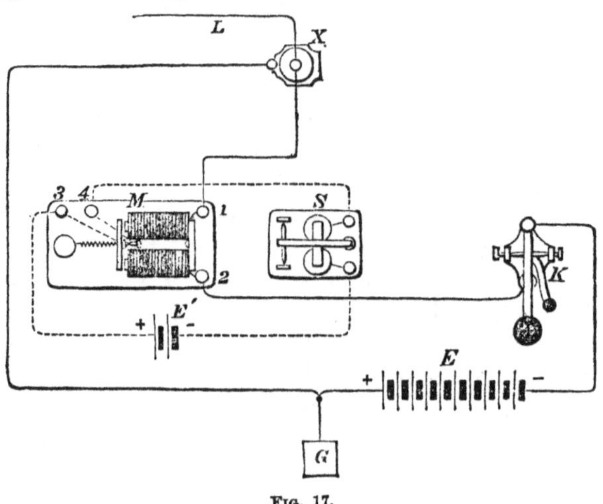

FIG. 17.

and instruments for one of the terminal stations of a

RELAY MAGNET.

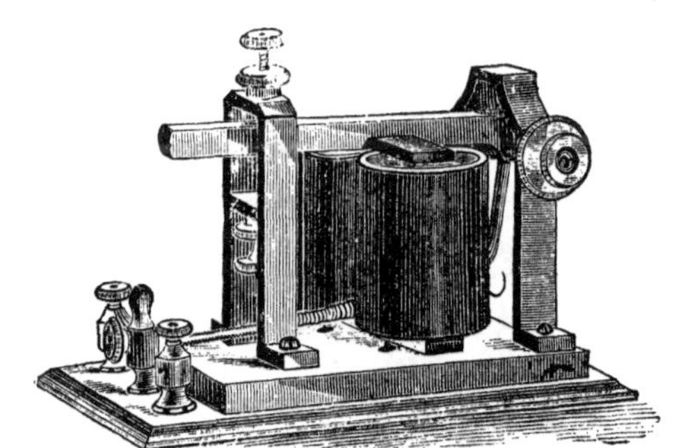

LOCAL SOUNDER.

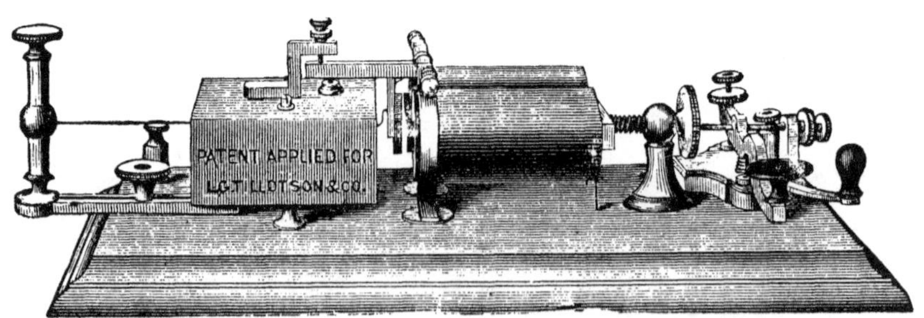

COMBINATION MAIN LINE INSTRUMENT.

Manufactured by L. G. Tillotson & Co., New York.

line. The line wire L first enters the lightning arrester X, and passes thence through the coils of the relay M by the binding screws, 1, 2, and thence to the key K, main battery E, and finally to the ground at G. The local circuit commences at the + pole of the local battery E′ and through the platina points of the relay by the binding screws, 3, 4, thence through the register or sounder coils, S, and back to the other pole of the battery.

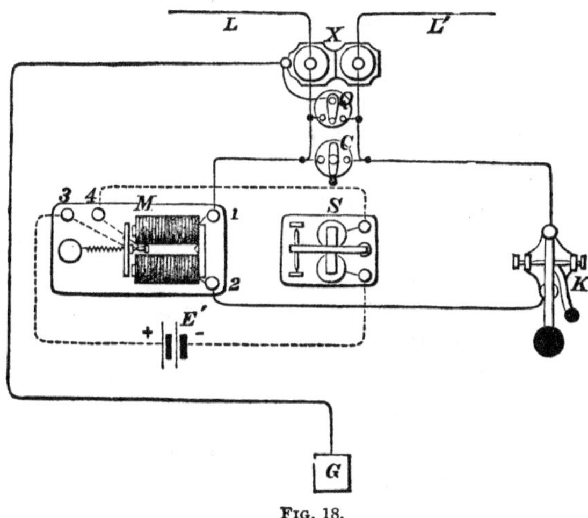

Fig. 18.

59. Arrangement of a Way Station.—Fig. 18 shows a plan of the instruments and connections at a way station. The line enters at L, passes through the lightning arrester X (70), and thence through the relay M, key K, and back to the lightning arrester, and thence to the next station by the line L′. The arrangement of the local circuit is the same as in the last figure. The button C, arranged as shown in the figure, is called a "*cut-out*" (62). When turned so as to connect the two wires leading into the office, it allows the line current to pass across from one to the other without going through the instruments. The instruments should always be cut out by means of this apparatus when leaving the office temporarily, or for the night, and

also during a thunder storm, to avoid damage to the apparatus. Fig. 21 shows a better arrangement.

The *Ground Switch*, Q (63), is used to connect the line with the earth on either side of the instruments at pleasure. It is only used in case of accidents or interruptions on the lines, as will be hereafter explained.

60. ADJUSTMENT OF THE APPARATUS.—The principal difficulties which the operator is liable to meet with in working the Morse apparatus are as follows :

1. When the paper in the register does not run freely from the reel on which it is held, or sticks in the guides from irregularity in width, or if the style is adjusted to indent the paper too deeply, the paper moves irregularly, shortening dashes into dots, and causing dots to run together.

2. The style should be adjusted so as to move freely in the groove of the upper roller, or the marks will be more or less indistinct. If it is completely out of the groove, no marks will be produced. These faults generally arise from too much end play in the pivots of the lever, or from the pivot screws working loose. When the lever works too loosely in its bearings, irregular dashes, too deep at their commencement, and tapering off to nothing, will be produced.

Residuary magnetism sometimes causes the armature of the electro-magnet to *stick*. This will always happen if the armature is allowed to touch the poles of the magnet. The screw stop should therefore be adjusted so as to prevent the armature from approaching too closely to the poles of the magnet. The upper screw stop, which regulates the play of the lever, should be adjusted so that the movement is just sufficient to withdraw the style from contact with the paper.

3. If the paper runs between the rollers "crooked," the pressure of the upper roller upon the paper is greater at one end than the other. This pressure is regulated by two springs, one on each side of the instrument, and they should be made as nearly equal in pressure as possible.

4. When the signs are confused the relay requires adjustment to suit the strength of the current.

5. If the relay moves by the action of the line current, and the register or sounder does not act, the fault is somewhere in the local circuit. If the register does not work when the relay is moved by the finger, the local circuit is certainly at fault, either from weakness of the local battery, a loose connection, a broken wire, or dirt between the platina points of the relay. The latter should, when too much corroded, be cleaned carefully with emery paper, taking care to remove as little of the platina as possible.

6. The *sticking* of the key, which sometimes occurs, is caused either by the platina points becoming oxidized and dirty, or by small particles of metal and dirt collecting behind the circuit closer and about the anvil, causing a partial connection when the key is open.

7. *It is very important that all the connections about an office should be firmly screwed up.* Neglect of this precaution is a *very prolific* cause of trouble upon a telegraph line.

8. In rainy weather, or when the insulation of the line is defective from any cause, the cores of the relay must be withdrawn to a greater distance from the armature, to avoid the influence of the residual magnetism, caused by the escape of the "current" from the line. This is called "adjusting" the instrument, and is one of the most important of an operator's duties, requiring great judgment and skill during unfavorable weather and on poorly insulated lines. The key should never be opened without carefully adjusting the relay, to be *sure* that no other offices are using the line.

SWITCHES OR COMMUTATORS.

61. These are employed for the purpose of connecting one circuit with another, for dividing a circuit into two parts, or in short, for any purpose where it is necessary to alter the connections of a line or circuit.

62. Fig. 19 shows the simple *Button* or *Circuit Closer*,

which is usually employed as a "cut out" (58). The base A is of wood or hard rubber. The brass lever, B, when in the position shown in the figure, forms an electrical connection between the metallic studs C C, which

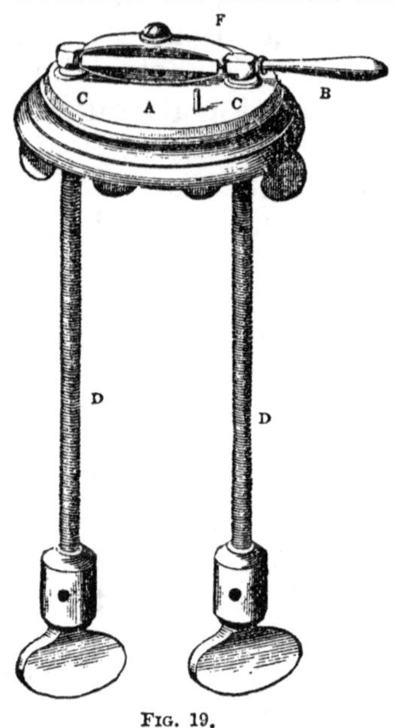

Fig. 19.

are continuous with the screws, D D, passing through the table and terminating in binding screws, to which the wires are attached. The spring F, pressing against the lever, insures a firm contact with the studs. This circuit closer is sometimes, for special purposes, made with four connections instead of two.

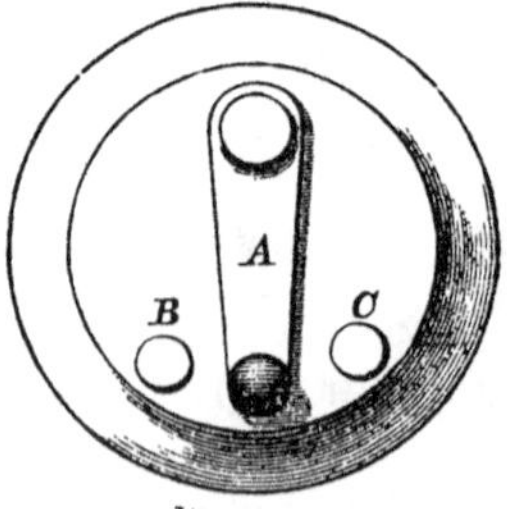

Fig. 20.

63. Fig. 20 represents a *Ground Switch* (58). The lever

A is attached to a wire leading to the earth, and the two studs, B, C, are connected to the line wire on each side of the instruments.

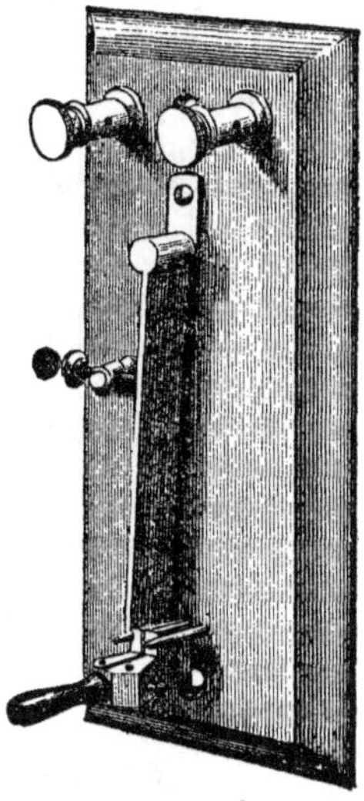

FIG. 21.

64. THE PLUG SWITCH is shown in fig. 21. This arrangement consists of a brass spring, brought very firmly against a stationary pin. A wedge or plug made of two pieces of brass, separated by an insulating material, is made in the form shown, to admit of insertion between the spring and the pin. The wires leading to the instrument are attached to this wedge by flexible conductors. When the wedge is inserted, the line current is diverted through the instrument, but is not interrupted. The instrument may readily be withdrawn from the line by taking out the wedge, the spring instantaneously closing the main circuit. This arrangement is found extremely useful in connecting batteries as well as instruments. At a way station it is preferable to a simple cut-out, for the reason that the apparatus is entirely disconnected from the circuit when the wedge is withdrawn (59).

65. THE UNIVERSAL SWITCH, for the use of offices having a considerable number of wires, is constructed

in several different forms, although the principle involved is nearly the same in each. Fig. 22* represents

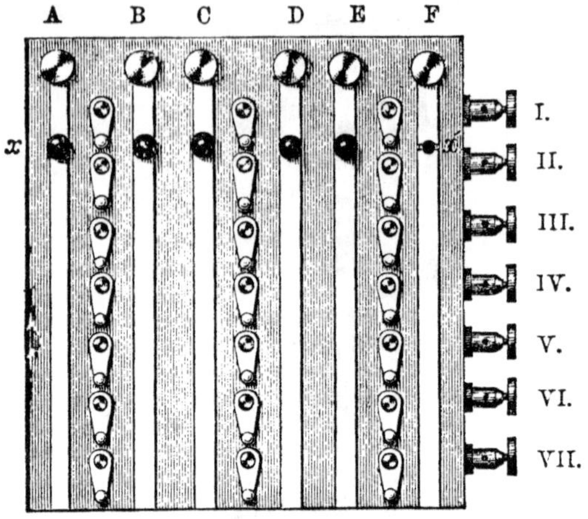

FIG. 22.

the arrangement most generally used, which is known as the Culgan Switch, from the name of its inventor. The upright straps of brass, A, B, C, D, E, F, are fixed upon a slab of hard wood, or other non-conducting material, and provided with binding screws at their upper extremities, for the reception of the line wires. The binding screws, I, II, III, IV, V, VI, are in electrical connection with the horizontal rows of buttons, by wires underneath the board, not shown in the figure. Thus, any wire attached to one set of binding screws may readily be connected with any wire attached to the other set, by simply turning the appropriate button. A row of metallic pegs, x x', are so arranged that either of the upright straps may be separated into two parts by the withdrawal of the peg belonging to it, as shown at x'. The object of this device will be explained hereafter.

This switch may be made of any size and with any number of connections, depending upon the number of lines it is designed to accommodate. The wires may be attached to it in a number of different ways, the parti-

* L. G. Tillotson & Co., New York.

cular arrangement adopted in each case depending upon the nature of the changes required to be made.

66. Arrangement of the Connections.—The switch shown in the figure, placed at a *way station*, could be arranged to accommodate three through wires, and an equal number of instruments, providing for all the necessary changes. The arrangements in this case would be as follows: Connect line wires Nos. 1, 2 and 3, *east*, with A, B and C; 1, 2 and 3, *west*, with D, E and F. Instrument No. 1 to I and II, No. 2 to III and IV, No. 3 to V and VI. Turn the buttons so as to connect A with I and D with II. The circuit of No. 1 wire will then enter at A, go to instrument No. 1 *via* I, returning to II, and thence going out at D. The other instruments may be connected at pleasure in the same manner. If it is desired to connect a circuit *through*, for instance No. 1, leaving the instrument out of circuit, it is done by turning the buttons so as to connect both A and D to the *same* horizontal wire, either I or II. By a little study it will be seen that any wire east may be connected with any other wire west, with or without any desired instrument, at pleasure. The ground wire is attached at VII, and may be connected with any line wire east or west at pleasure.

67. The same switch, placed at a *terminal station*, would provide for six wires, by connecting them as before to the screws A, B, C, D, E, F, and the instruments to I, II, III, IV, V, VI. The wires of a *loop* (87) may be connected to I and II in place of the instrument, and may be put in circuit with any wire by turning the buttons connected with I and II both on to the corresponding strap, which is then divided by withdrawing the peg, forcing the current to pass through the loop. Extra sets of buttons for loops are usually provided when the switch is intended for a terminal station, which can be used without diminishing the capacity of the switch for other purposes.

68. Jones' Lock Switch is employed for the same purposes, and connected in the same manner as the one

last described, but the connection between the vertical and horizontal wires is made by a metallic peg, provided with a spring, as shown in fig. 23 (*Chester*). This arrangement entirely obviates the danger of imperfect connections, from the loosening of buttons, etc, which is sometimes a source of trouble in the Culgan Switch.

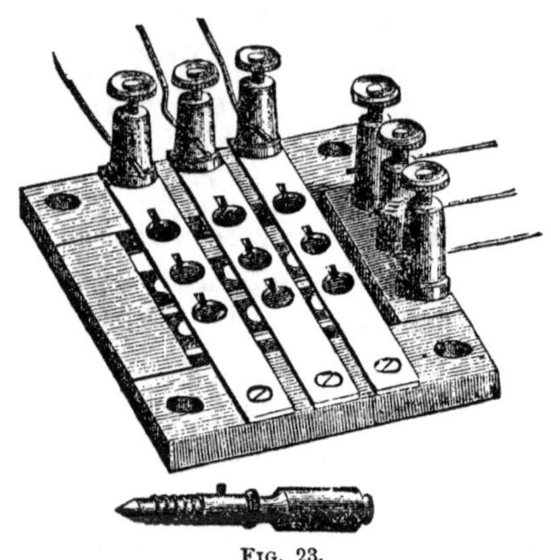

Fig. 23.

It is also cheaper and much more compact; a matter of some importance in arranging for the accommodation of a large number of wires.

69. There are other forms of switches designed for special purposes, which it is unnecessary to describe in a work of this kind. Those already referred to are all that are generally required in fitting up a telegraph station.

LIGHTNING ARRESTERS.

70. The danger of injury to the instruments and operators at a telegraph station, by atmospheric electricity, is usually guarded against by the use of an apparatus termed the *Lightning Arrester*, which is constructed in accordance with the well established fact that this kind of electricity, being possessed of enormous intensity, prefers a short route through a poor conductor to a longer one through a good conductor, while the comparatively low intensity of the voltaic

current, used for telegraphic purposes, confines it to the conducting wires.

71. THE PLATE ARRESTER.—The arrester most usually employed upon the telegraph lines in this country consists of a flat plate of brass, about five or six inches in length, which is attached to the "ground wire." Other plates of brass rest upon this, being separated from it by a thin sheet of insulating material. These last mentioned plates are provided with binding screws, for the attachment of the line wires. Any surplus charge of atmospheric electricity, entering by the line wires, forces its way through the insulating material into the ground plate, and is thus carried off to the ground without injuring the apparatus. The form of arrester supplied by the Messrs. Chester is shown in fig. 24.

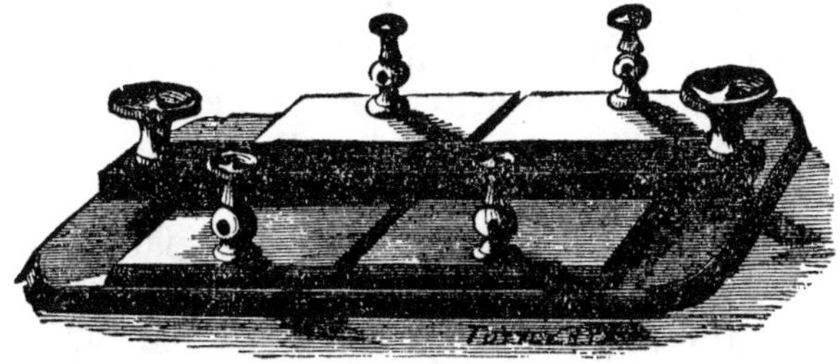

FIG. 24.

The plates in connection with the line wires are firmly held in their places by a wooden cross piece, secured by screws at each end, as shown in the cut. A thin sheet of gutta percha, or paper, is used to separate the plates. When paper is used it should be saturated with paraffine. Mica is, perhaps, better than either, as it is not carbonized by the passage of the spark, as paper sometimes is, so as to form a ground connection. The manner in which the arrester is connected with the wires leading into an office will be seen by reference to fig. 18, where the two line wires, L and L′, are attached to the two upper plates of the arrester, X, while a wire leading to the ground at G is attached to the lower plate.

72. BRADLEY'S ARRESTER.—Another form of arrester,

designed by Dr. Bradley, is shown in fig. 25, and has recently been quite extensively employed, with excellent results.

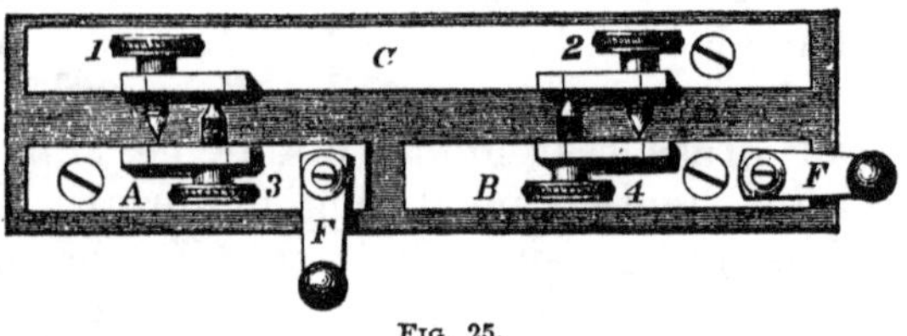

FIG. 25.

It depends for its action upon the well ascertained fact that lightning always passes from a point to a plate with great facility. The line wires leading into the office are attached to the metallic plates A and B by means of binding screws beneath, the ground wire being attached in the same manner to the plate C. Platina tipped screws, 1, 2, 3, 4, are fixed to each plate, and are adjusted so as to come nearly in contact with the opposite plate. As lightning occasionally passes from the earth to the clouds, as well as from the clouds to the earth, this arrester is so arranged as to facilitate its passage in either direction. The buttons, F F, are so arranged that the apparatus serves for a "cut-out" and a "ground switch" as well as an arrester. Its application to these purposes will be at once understood by an inspection of the cut. This form of arrester is peculiarly well adapted for the protection of cables, or any situation where it is exposed to accidental dampness, as it is much less apt to interfere with the working of the line in such cases than the plate arrester previously described.

73. Lightning arresters must always be kept free from dampness and dirt, as far as practicable. Much annoyance often arises from neglect of this precaution, as moisture between the plates will often cause a serious escape, greatly interfering with the working of the line. This difficulty is especially liable to occur where the arresters are used for the protection of submarine cables. A flash of atmospheric electricity also fre-

quently carbonizes the paper between the plates, or fuses the metal, so as to permanently connect the ground and the line. Consequently, the lightning arresters should be frequently taken apart and examined. *This should invariably be done after a thunder storm.*

REPEATERS.

74. When the length of a telegraphic circuit exceeds a certain limit, depending upon the insulation, the size of the conductor, the number of instruments in circuit, etc., the line current becomes so enfeebled, even when large batteries are employed, that satisfactory signals cannot be transmitted. In such cases it was formerly customary to re-write the messages at some intermediate station, but this duty is now usually performed by an apparatus called a *repeater*. The principle of this arrangement consists in causing the sounder or register connected with one circuit to open and close the circuit of another line by an action similar to that of a relay (56). Repeaters are also often used for connecting one or more branch lines with a main line, for the purpose of transmitting press news, etc., simultaneously to different places. This enables all the stations in connection to write to each other as readily as if they were situated upon the same circuit.

Since the general introduction of repeaters it has become quite practicable to telegraph direct between places situated at very great distances from each other. It is not uncommon, at the present day, to work direct through four or five thousand miles of continuous line by the aid of these instruments with almost as much facility as if it were one continuous circuit. On one or two occasions the stations at Heart's Content, Newfoundland, and San Francisco, California, have been placed in direct communication with each other, the operators at these widely separated points conversing with each other across the entire breadth of the continent without the slightest difficulty.

75. WOOD'S BUTTON REPEATER.—This is the simplest arrangement of this kind now in use. Fig. 26 shows

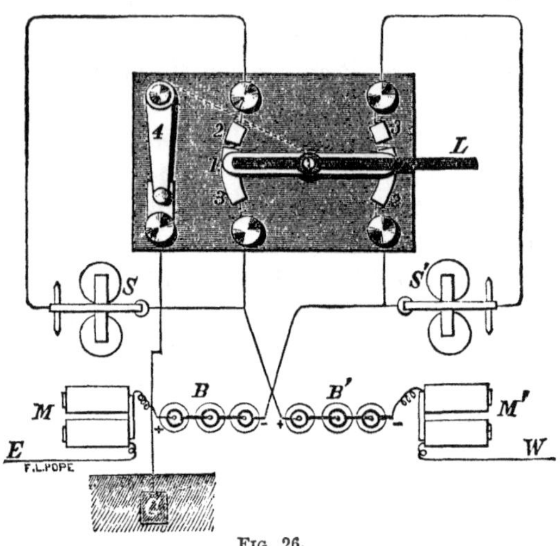

FIG. 26.

the most convenient and serviceable form in which the button or switch, and its connections, can be arranged for the purpose of changing the circuits. The instruments, batteries, &c., are shown in outline, for convenience of explanation. M and M′ are the eastern and western relays, S and S′ the eastern and western sounders. The local connections are not shown, but are run as usual. The eastern and western main batteries are shown at B and B′, and are placed with *opposite* poles to the ground, at the repeating station, so that when the line is put "through" the two batteries will coincide.

By means of this arrangement the following result may be obtained :

I. *Two distinct and independent circuits.* The lever L remaining in the position shown in the drawing (marked 1), and the button at 4, *closed.*

II. A *through circuit.* The lever L remains as before, but the button at 4 is *opened,* throwing off the ground connection between the two batteries, B and B′.

III. *Two distinct circuits arranged for repeating.* The button at 4 is closed. If the lever L be placed in the position indicated by the figures 2, 2, the eastern sounder repeats into the western circuit. If the lever is changed to 3, 3, the western sounder repeats into the eastern circuit. The operator in charge of a button repeater will find his duty very simple if he governs himself by the following

RULE.—When either sounder fails to work coincident with the other, *turn the button instantly*.

In connecting up this apparatus, the arrangement of the poles of the main batteries above specified should be carefully borne in mind. It is also of the utmost importance that these batteries should be *perfectly insulated* from the ground, as the point at which the circuit is open and closed is between the battery and the ground. Therefore, an escape occurring from the battery to the ground will cause a residual current upon the main line, when the circuit is open at the repeating points of the sounder, and thus interfere with its working.

In cases where it is not required to work the two lines through in one circuit, the connections are arranged differently from the plan shown in fig. 26, the main battery being placed in the circuit between the lever L and the ground G, instead of at B and B′, as shown. In this case the switch 4 may be dispensed with altogether.

76. The lever of the sounder moves through a certain space before closing the circuit of the second line, so that the duration of the current sent forward is shorter than that received from the transmitting station. A second repeater shortens it still more, so that the dots cease to be repeated, and are frequently lost altogether. The sending operator must therefore transmit the signals more *firmly*, as it is termed ; that is, increase the length of the key contact, especially when sending dots. For the same reason, the sounder levers in a repeating apparatus should be adjusted to have as little motion as possible.

77. HICKS' AUTOMATIC REPEATER.—This arrangement

dispenses with the attendance of an operator for the purpose of changing the circuits while working, the only attention required being to keep the relays properly adjusted. The principle of the apparatus is shown in fig. 27.

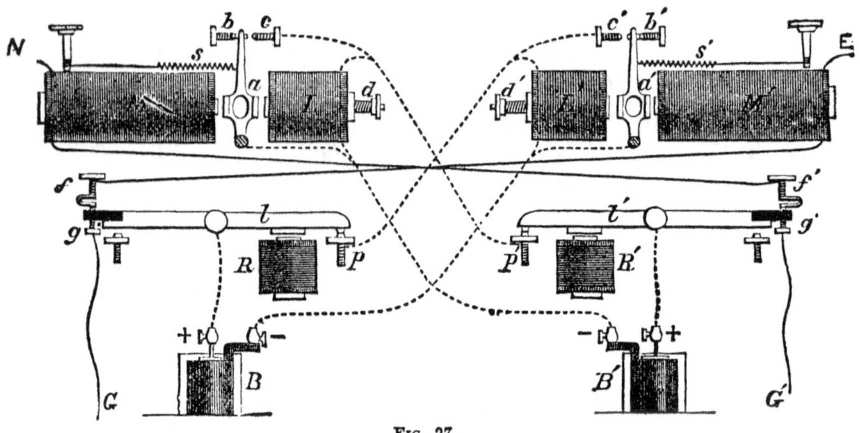

Fig. 27.

The main circuits pass through the relay magnets M and M′, thence to the repeating points *f g* and *f′ g′*, attached to the opposite sounder levers respectively, and thence to the main battery and ground at G and G′. The platina points of the screws *f* and *f′* are placed upon U shaped springs, which, in a great measure, prevents the shortening of the signals referred to in the last paragraph. The local circuits are run through the relay points *b* and *b′* and the sounders R and R′, on each side of the apparatus, in the ordinary manner, but to prevent confusion of lines, are omitted in the drawing. The "extra local" magnets, L and L′ act upon armatures placed upon the relay levers *a* and *a′*, opposite to the regular armature. (See figure.) These extra local magnets are movable by means of the screws *d d′*, and the adjustment of the relays M M′ is performed by means of these extra local magnets, the springs *s s′* not being used for this purpose.

In the figure the repeater is shown in its normal position, with both circuits closed. The circuits of the extra local batteries B B′ (shown by dotted lines) pass through the sounder levers *l l′*, the screws *p p′*, and thence respectively to the extra local magnets on the opposite side of the apparatus. These magnets must be so adjusted that their attraction is not sufficient to draw the armatures away from M M′ unless the main circuit is broken.

It will also be seen, by referring to the drawing, that when the main circuit is broken and the armature falls back on the point *c*, that the extra local magnet L is *cut out*. But the instant this happens the spring *s* draws the armature away again. As soon as the contact is broken at *c* there is a circuit through L, and the armature is again drawn back to *c*. The tension of the spring *s* being but just sufficient to draw the armature away from *c*, the armature *vibrates* on the point *c* through such a small space, and with such rapidity, that the motion is invisible to the eye. On account of the extreme rapidity of these vibrations, it is impossible to close the main circuit at a time when the extra local magnet L is *not cut out*, and the armature will consequently obey the slightest impulse caused by the attraction of the relay magnet.

The working of the apparatus requires but little further explanation. If the western main circuit be broken, for instance, the armature lever *a* falls back and vibrates on the point *c*, as above described. The sounder lever *l* first breaks the circuit of the eastern extra local magnet L′, then that of the eastern main line, which passes through the relay M. The circuit through both L′ and M′ being thus broken, the slight tension of the spring s′ will hold the armature in its place, and prevent the local circuit through R, and consequently the western main circuit, from being broken. When the western circuit is again closed the reverse of these operations takes place.

78. In using this repeater the springs *s s′* should be adjusted with the smallest possible amount of tension, just sufficient to hold the armature in place. *When once*

adjusted they should be let alone. Care must be taken that none of the wires under or about the magnets touch any part of the brass. The extra local magnets, for example, may be cut out entirely in this way. The screws that adjust the extra local magnets should be oiled with fine oil to prevent wear and make their adjustment easy. The extra local batteries must be kept of a uniform strength ; if they are allowed to become weak the instrument will be thrown out of adjustment.

79. MILLIKEN'S REPEATER.—In the general arrangement of its connections this repeater somewhat resembles that of Hicks', but is more simple in principle. Fig. 28 is a plan of its connections. The main line

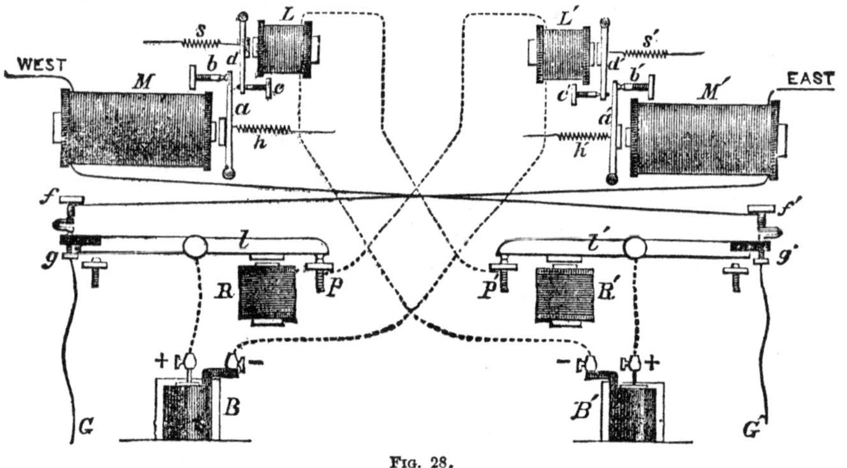

FIG. 28.

wire from the west passes through the relay magnet M and the repeating points f' g' of the opposite sounder, and thence to the battery and ground at G′. The eastern line passes through M′, f and g to G, in a similar manner.

The extra local magnets L and L′ are arranged, as shown in the figure, so that when either of their armatures is released it is drawn back by the spring attached to its lever, bringing the latter firmly in contact with the armature lever of the corresponding relay. The extra local batteries are shown at B and B′ the circuit of each being indicated by dotted lines. The ordinary local circuit through the relay and sounder is omitted, to avoid confusion in the diagram.

If the main circuit be broken in the western wire, the relay M breaks the local circuit of the sounder R at *b*. The movement of the lever *l* of the sounder first breaks the extra local circuit at *p*, causing the magnet L′ to release the armature *d*′, which is drawn back by the spring *s*′ against the top of the lever *a*′, and, secondly, the eastern main circuit is also broken at *f*. *g*. The lever *a*′ is prevented from falling back when the circuit of M′ is broken by the tension of the spring *s*′, which is so adjusted as to be greater than that of the spring *h*′. The apparatus on the right hand side of the repeater, therefore, remains quiet while the west is working, and *vice versa*, the current through M′ being always restored before that through L′ is broken, which is effected by the U shaped spring on the screw *f*.

One of the principal advantages in the construction of Milliken's repeater consists in the fact, that any slight variation in the strength of the extra local circuit, from weakness of the battery or other causes, does not affect the adjustment of the relay magnets, as in the case with Hicks' repeater. The adjustment and action of the two magnets are entirely independent of each other, as will be seen by reference to the diagram. The relay levers also move more freely, being unencumbered with extra armatures or other appliances.

In this, as in the Hicks repeater, buttons are provided, by means of which each line may be worked separately without interfering with the other, if desired.

These are omitted in the drawing, to prevent confusion, but are arranged so that, when closed, one button forms a permanent connection between *f* and *g*, thus preventing the movement of the lever *l* from breaking the eastern main circuit, and another connects *p* and *l*, thus keeping the extra local circuit constantly closed, and the armature lever *d*′ withdrawn from interference with *a*′.

The same thing may be accomplished by causing the button to break the extra local circuit entirely, when the instruments are to be worked separately, and "turning down" the adjusting spring *s'* of the lever *d'*. It will, of course, be understood that the other side of the repeater is arranged in precisely the same manner.

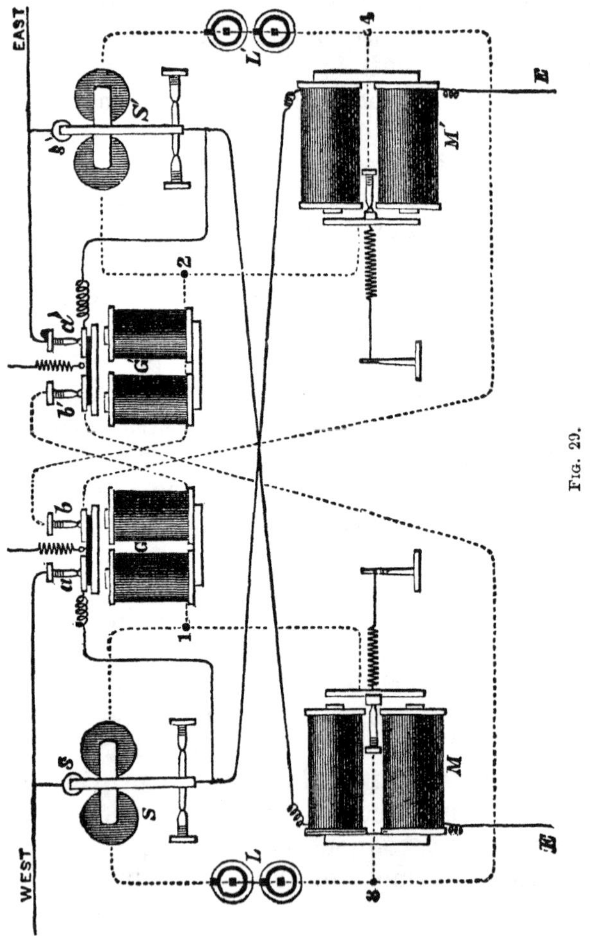

Fig. 29.

80. Bunnell's Repeater.—The arrangement of the main circuits in this repeater is exactly the same as in

the ordinary "button repeater," and will be readily understood by reference to fig. 29. The eastern main wire enters at the right, passing through the repeating point, *s*′, of the western sounder, S′, and through the coils of the eastern relay, M, and thence to the main battery and earth at E. The western main wire is similarly connected on the opposite side of the instrument. In the button repeater (75) a switch is so arranged as to form a connection, cutting out the repeating points of the sounder on the opposite side, when either line is working, requiring a person to be constantly stationed at the instrument to make the necessary changes when two stations, on opposite sides of the repeater, are corresponding with each other. In Bunnell's repeater this duty is performed automatically by means of two "governor" or controlling magnets, G G, the action of which will be hereafter described.

The eastern and western main circuits both being closed and the apparatus at rest, the course of the local circuit of the eastern instrument is as follows: From the local battery, L, through the coils of the eastern sounder, thence passing through the closed relay points at M, and returning to the other pole of the battery. The resistance of the governor magnet, G, prevents any appreciable portion of the current from passing through its coils, as long as the closed points of the relay, M, afford it a shorter route. If the local circuit be broken by the relay points at M, it is forced to pass through the coils of the sounder, S, and also of the governor, G.

When a circuit of low intensity passes through the coils of two magnets, differing considerably in resistance, the attraction of the magnet having the least resistance is very small in comparison with that of the other. A practical application of this principle is made in this repeater, by forming the helices of the governor magnet of finer wire than that of the sounders. The effect of this is, that when the local circuit is thrown through both magnets by the opening of the relay, that the armature of the governor magnet is attracted with considerable strength, while the magnetism developed in the sounder is not sufficient to move its armature, although the same current passes through its coils. This arrangement is, of course, the same on each side of the repeater, and by bearing it in mind the action of the instrument may be readily comprehended.

When both main circuits are closed and the repeater at rest, the governor magnets remain open, being cut out by the points of the relays, which, as well as the sounders, are closed on both sides of the apparatus. If, now, we suppose the circuit to be opened by an operator on the western main line, the armature of the relay, M′, falls back, opening the sounder, S′, and closing the governor magnet, G′, as previously explained. This breaks the eastern main circuit at *s*′, and also at *a*′, as well as the circuit of the opposite governor magnet, G, at the point *b*. The breaking of the eastern main circuit at S′ opens the eastern relay, M, and consequently its sounder, S, but the circuit of the governor magnet, G, being broken at *b*′, it remains inactive, and the western main circuit is complete through the points, *a*, although broken at the point, *s*, by the opening of the sounder, S. Upon the closing of the western main circuit this action is reversed, and the apparatus resumes its original position. If the eastern main circuit be opened the same action takes place, but on the opposite side of the repeater.

In most repeaters hitherto constructed one side of the apparatus remains silent while the opposite side is in action, but in this one the relays and sounders on both sides work together, the points, *a*, *a*′, on the armature of the governor magnets acting automatically in the same manner as the switch of a button repeater, when moved by the hand of the operator.

81. An advantage claimed for this repeater is, that both sides of the apparatus work together, affording the operator in charge a better opportunity to know how both lines are working. The extra local batteries are dispensed with, and the relay levers are not encumbered

with extra armatures and other appliances. The adjustments required are the same as in a simple relay and sounder.

82. Various other repeaters have been contrived, and to some extent adopted in this country, but as those we have described are much more extensively used than any others, it has not been deemed necessary to describe the others in a work of this kind.

83. COMBINATION LOCALS.—In offices containing a number of instruments, a single local battery is frequently employed to operate all the sounders in the office. Such an arrangement is called a combination local. The best way of making the connections is shown in fig. 30, in which the instruments are represented at

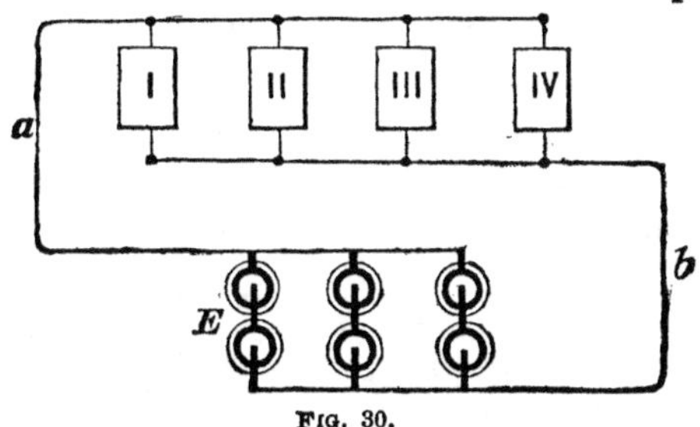

FIG. 30.

I, II, III and IV. The local battery is shown at E. The common conductors, *a* and *b*, should be of large copper wire, say No. 12 or 14. If the ordinary Daniell's battery is used for this purpose, the cells should be connected for *quantity*, as shown in the diagram, and not in a single series. Every sounder in the combination should have the same size and amount of wire in its coils, as nearly as possible, in order to secure the best results.

84. Another plan is to use separate locals, a wire being run from one pole of each local to its corresponding instrument, the opposite poles of the batteries, and the instrument wires being all connected to a common return wire.

85. These combination locals are very objectionable, however, and their use should be avoided wherever

possible. The iron cores in two different relays may happen to be in connection with the silk covered wire with which they are wound, a circumstance which frequently occurs. In such a case, if the two armatures chance to touch the poles of their respective relays, a metallic connection, technically called a *cross*, is made between the two main lines. Again, if these two relays are at a terminal station, and in connection with two main batteries, with opposite poles to the ground, the combined force of both batteries is thrown on short circuit, through the local return wire, burning the relays, exhausting the batteries, and interfering with the operation of every wire connected with them. The cause of these troubles being somewhat obscure, it might, for a considerable time, escape detection.

86. Local Circuit Changer.—In offices containing two sets of instruments on different circuits, it is often desirable to change them. A simple arrangement for this purpose it shown in fig. 31, in which the relays are

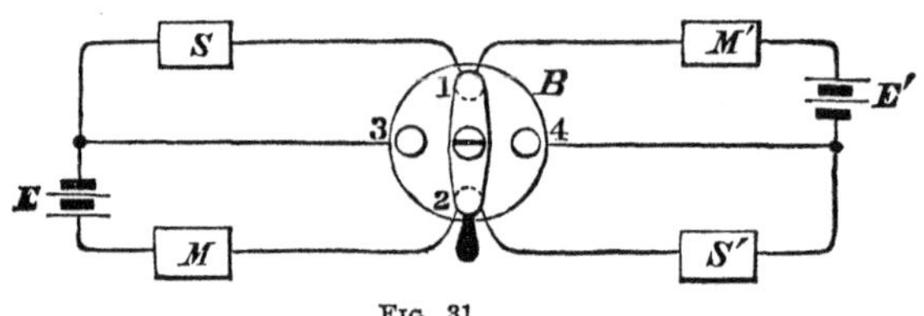

Fig. 31.

represented at M and M′; S and S′ are the sounders or registers, E and E′ are the local batteries. B is a simple button or circuit closer (62), having four connecting points, 1, 2, 3, 4. When the button is in the position 1, 2, as shown in the figure, the relay M works the sounder S, and the relay M′ the Sounder S′. By changing the position of the button to 3, 4, S is worked by M′ and S′ by M. This simple arrangement is often very convenient in railway stations, where a sounder may be placed on one circuit and a register on the other, so that an operator who is unable to read by sound can instantly shift the register upon either line at pleasure.

TECHNICAL TERMS USED IN THE TELEGRAPH SERVICE.

87. *Line.*—The wire or wires connecting one station with another.

Circuit.—The wires, instruments, &c., through which the current passes from one pole of the battery to the other.

Metallic Circuit.—A circuit in which a return wire is used in place of the earth.

Local Circuit.—One which includes only the apparatus in an office, and is closed by a relay.

Local.—The battery of a local circuit.

Loop.—A wire going out and returning to the same point, as to a branch office, and forming part of a main circuit.

Binding Screws or Terminals.—Screws attached to instruments for holding the connecting wires.

To Cross-connect Wires.—To interchange them at an intermediate station, as in § 117.

To put Wires straight.—To restore the usual arrangement of wires and instruments.

To Ground a Wire, or put on Ground.—To make a connection between the line wire and the earth.

To Open a Wire.—To disconnect it so that no current can pass.

Reversed Batteries.—Two batteries in the same circuit with like poles towards each other.

To Reverse a Battery.—To place its opposite pole to the line; or, in other words, interchange the ground and line wires at the poles of the battery.

Escape.—The leakage of current from the line to the ground, caused by defective insulation and contact with partial conductors.

Cross.—A metallic connection between two wires, arising from their coming in contact with each other, or from other causes.

Weather Cross.—The leakage of current from one wire to another during rainy weather, owing to defective insulation.

CHAPTER V.

INSULATION.

88. A telegraph wire suspended on poles is attached to *insulators*, to prevent the escape of the current to the earth at the points of support. Insulators should be regarded in the light of *conductors*, whose value depends upon their resistance to the passage of the current.

89. The insulation of a line is never perfect, even in the dryest weather. There is a leakage at every support, which is greatly increased when the surfaces of the insulators are damp, especially if covered with smoke or dirt. Experiments show that soot will destroy the surface insulation of the best insulators, even when exposed to the cleansing action of the rain. This evil is confined, however, principally to cities, and does not manifest itself to nearly so great an extent in the open country.

90. Insulators, considered as conductors, follow the same law as other conductors. The less the diameter and the greater the length, the more resistance is opposed to the escape of the current. As in this case the resistance is almost entirely a question of surface, the best insulator is that having the smallest diameter and the greatest length between the wire and the support. The latter is accomplished by making the insulator of a cup form, or still better, of two cups, one placed within the other.

91. The material of which the insulator is composed should be a poor conductor of electricity and heat, a non-absorbent of moisture, with a surface repellant of water, and free from pores or cracks. It should also remain unaffected by exposure to the weather, and the effects of heat and cold. Nearly all of the materials

ordinarily employed are, however, liable to some of these objections.

Insulators of glass and porcelain being conductors of heat, a change of temperature from cold to warm causes a condensation of moisture upon their surfaces, including the portion protected from the direct action of rain, and from this arises the principal objection to the use of these substances in the construction of an insulator.

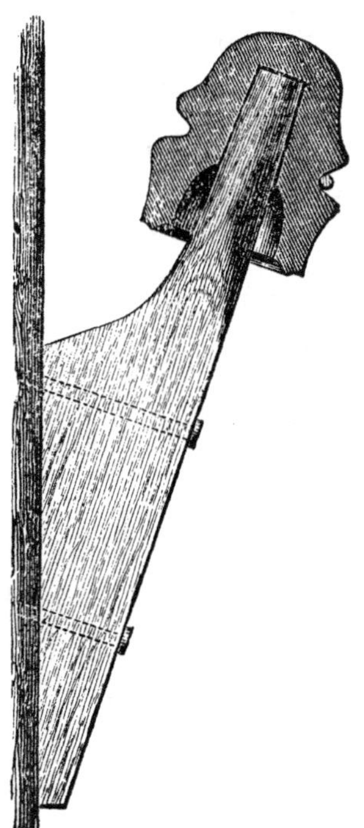

FIG. 32.

Hard rubber is in itself a better insulator than glass; but its surface, from exposure to atmospheric influences, soon loses its property of repelling moisture, and becomes rough and porous.

A surface which repels watery accumulations will cause them to flow disconnectedly in drops, instead of forming a continuous conducting film. This property is therefore one of great value for the purposes under consideration.

92. THE GLASS INSULATOR.—The insulator most commonly employed in this country is the glass. This is generally made in the form represented by Fig. 32, which is a sectional view of the insulator fixed upon a wooden bracket, the latter being securely spiked to the side of the pole. The line wire passes alongside the groove surrounding the insulator, and is fastened with a *tie-wire* encircling the insulator, both ends of which are wrapped around the line wire. The concavity of the under side of the glass keeps it dry, in some measure preventing the current from escaping to the wet

bracket and pole through the medium of a continuous stream of water.

93. The Wade Insulator.—This is largely used in the Western States. Its construction is shown in Fig. 33.

Fig. 33.

A glass insulator, somewhat similar in shape to that last described, is covered with a wooden shield, to prevent fracture from stones and other causes, the wood being thoroughly saturated with hot coal tar, to preserve it from decay. The line wire is tied to the outside of the shield, in the same manner as when the glass insulator is used.

This insulator is usually mounted upon an oak bracket, as in Fig. 33, secured by spikes to the side of the pole or other support. When it is intended to be mounted upon a horizontal cross-arm it is placed upon a straight wooden pin, instead of a bracket. The pin or bracket is usually saturated with hot coal tar, in the same manner as the insulator shield.

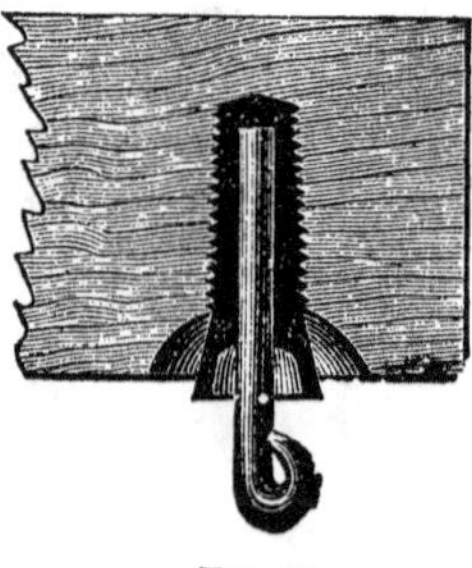

Fig. 34.

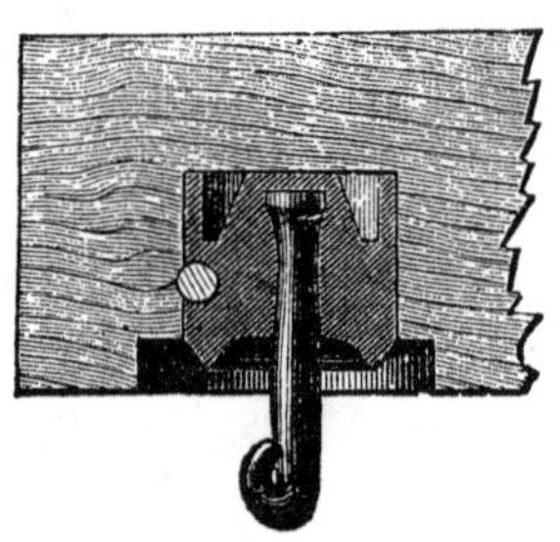

Fig. 35.

94. Farmer's Hard Rubber Insulator.—This is shown in Fig. 34. It is a good insulator when new,

but by exposure to the weather its surface becomes rough and spongy, and retentive of moisture. It is screwed to the under side of the cross-arm or wooden block, which is secured to the pole. The best form is that which is made with a drip or shed, as shown in the figure. If exposed to the direct action of rain it ought always to be placed in a perpendicular position. It will be noticed that this insulator holds the line wire by suspension.

95. THE LEFFERTS INSULATOR.—This is composed of a suspension hook fixed in a socket of glass, of the form represented in Fig. 35. This is inserted into a hole bored in the under side of a block or cross-arm, and fastened with a wooden pin. In painting the arm or blocks the paint must not be allowed to get on the surface of the glass.

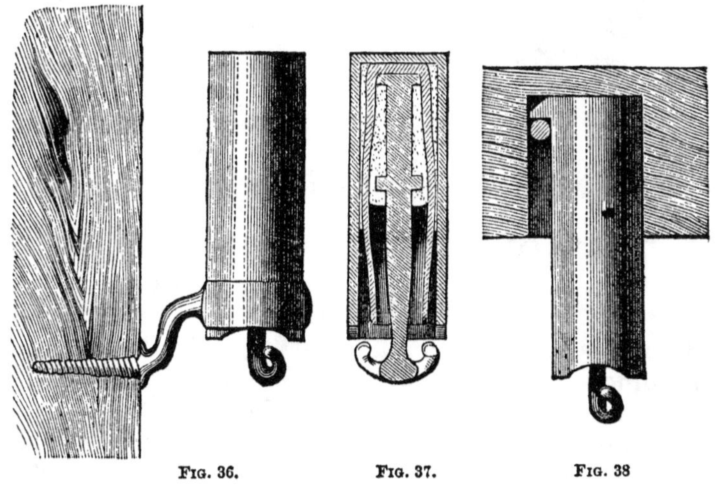

FIG. 36. FIG. 37. FIG. 38

96. THE BROOKS INSULATOR.—Figs. 36 and 37 show the construction of this insulator, which consists of a suspension hook cemented into an inverted blown glass bottle, which is again cemented into a cast iron shell, provided with an arm which screws into the pole, as in Fig. 36. Another form is made, designed for attachment

to a cross-arm, as in Fig. 38. The remarkable insulating properties of this arrangement are mostly due to the use of paraffine, with which the cementing material (sulphur) is saturated. It has also been discovered that blown glass possesses extraordinary properties of repelling moisture. Additional advantage of this fact has been taken in the construction of this insulator, as may be seen by reference to the cut.

97. Some important improvements have quite recently been made in the mechanical construction of the Brooks insulator, which are shown in Fig. 39. In the old form of hook, shown in Fig. 37, the wire has three bearings. To hold the wire securely, it is necessary that these bearings should be so direct as to make it difficult to place the wire in it, and the latter is often weakened by being bent. The new hook, shown in Fig. 39, has five bearings for the wire, but not so direct as to injure or weaken it by bending. The wire can be placed in this hook without labor or difficulty, and a strain cannot be applied in any direction by means of which the wire can be removed or released.

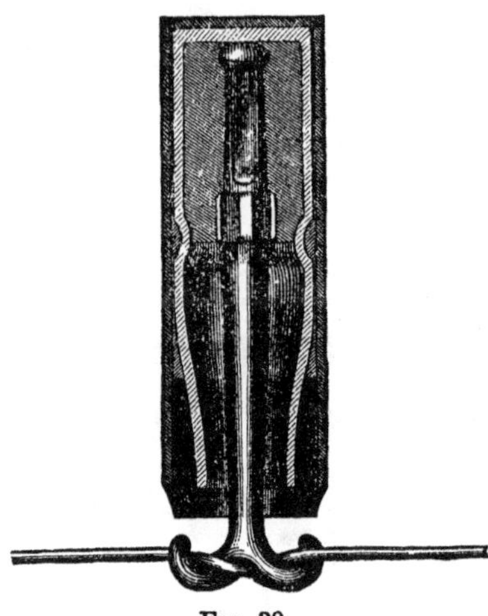

Fig. 39.

98. Mode of Testing Insulators.—The proper way to test the comparative value of insulators is to fix them upon frames or standards, in sets of ten or more, and place them where they will be fully exposed to the weather. The tests should be made when the weather is very wet, by means of a wire attached to all of them in the usual manner, and leading to the testing instrument, battery and ground. By this means the relative resistances of either of the insulators above described, and their consequent value in the construction of a line, may be readily ascertained.

99. Escape.—When the insulation is defective, or the wire comes in contact with the branches of trees, a wet wall, or other partial conductor, a portion of the current passes to the ground, forming what is technically known as an *escape*.

100. Weather Cross.—The escape of the current from one wire to another one upon the same poles, owing to defective insulation, is sometimes wrongly called "induction," or "sympathetic currents." Weather cross is a much more appropriate term.

As electric currents always move in the direction of the least resistance, their tendency is to escape from a long circuit to a shorter one. This mixing of the currents from different wires is a much more serious evil than a simple escape to ground, for the latter may in most cases be overcome by increased battery power; but when cross connection exists between different wires upon the same poles, an increase of battery upon one wire gives it an advantage over the others, but necessarily at their expense.

The effects of weather crosses usually manifest themselves upon the occurrence of a shower sooner than the escape to ground, because the horizontal arms become wet sooner than the vertical pole.

On the English lines this difficulty is obviated by means of an earth wire attached to each pole, and wrapped around the center of the arms, thus cutting off the currents passing from wire to wire, and conveying them to the ground. The battery can then be increased at will on one wire, without interference with the others. A much more economical and effective method of obtaining this result is that of improving the insulation.

101. Effect of Escapes and Grounds upon the Circuit.—If the wire touches a conductor communicating with the earth, or the earth itself, in a moist or wet place, so that the point of contact offers little or no resistance compared with the wire beyond, the fault is called a *ground*. The effect of a ground or escape is to increase the strength of the current going out to the

line, and to exhaust the batteries more rapidly. Therefore, in working with a continuous current, as is the case on American lines, the line current *increases* in strength in wet weather, but the *variation* or difference in the current at one station, when the line is opened and closed at another, *decreases*, and the effective signals are therefore weakened.

102. THE LAWS OF THE ELECTRIC CURRENT.—The laws which govern the propagation and distribution of electric currents are so simple, and at the same time so important, that every telegrapher should be familiar with them. By their aid the phenomena above referred to may be readily comprehended. The most important of these laws was first enunciated by Ohm, in 1827, and is known as *Ohm's law*. It may be briefly stated as follows:

Call the sum of the electro-motive forces...E
" " internal resistance of the battery..R
" " resistance of line and instruments..L
" " the effective strength of current...C

$$\text{Then } C = \frac{E}{R + L}$$

That is: *The effective strength of the electric current in any given circuit is equal to the sum of the electro-motive forces divided by the sum of the resistances* (174).

103. PRACTICAL APPLICATION OF OHM'S LAW.—FIRST CASE.—To illustrate the application of this law to circumstances occurring in practical telegraphy, take the

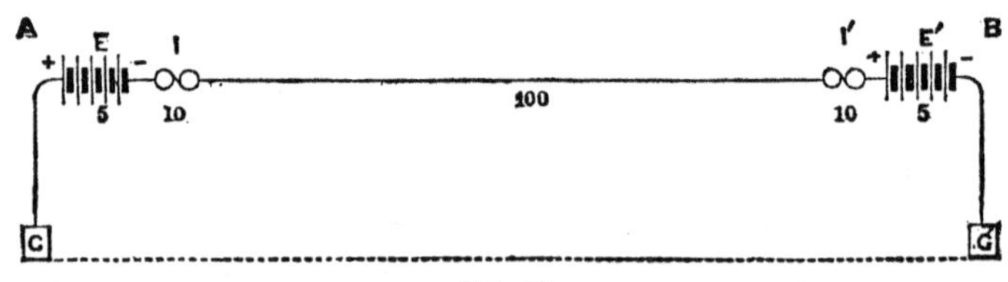

FIG. 40.

case of an ordinary telegraph line (Fig. 40), extending from A to B, and perfectly insulated, having a resistance of 100 Ohms. Let the main batteries, E and E' have each an electro-motive force of 1,000, and a resistance of 5 ohms, and let the resistance of the instruments I and I′ be equal to 10 ohms each. The total resistance of such a circuit will be :

100	ohms,	line,	} = L
20	"	instruments,	}
10	"	batteries,	= R
130	"	= R + L	

The line being perfectly insulated, the whole current from the batteries will necessarily act upon both instruments.

As the effective strength of the current in any circuit is, by Ohm's law, equal to $\frac{E}{R+L}$, in this case it will be

$\frac{2000}{130}$	= 15.4
With key open at A or B.............	= 00.0
Difference, or effective working strength.	= 15.4

If, on the above line, an escape occurs between the stations A and B, offering a resistance of 50 ohms, the effect will be the same as if a wire having a resistance of 50 ohms were connected from the centre of the line to the ground. The current from each battery has a tendency to divide at the fault between the two routes open to it, in proportion to their relative conductivity; or, what is the same thing, in inverse ratio to their respective resistances. But in this case the electromotive forces and the resistances are exactly the same on each side of the fault; and the positive current from one battery, and the negative from the other, have an equal tendency to escape to ground at the fault. These opposite tendencies consequently neutralize each other, and no effect whatever is produced upon the circuit by the fault as long as the line remains closed both at A and B.

If, however, A is sending to B, his key is alternately open and closed. When open, the circuit of the battery E (Fig. 41) is entirely broken. There will still, however, be a circuit from the battery E′, through I′ and the line to the fault F, and thence to the ground.

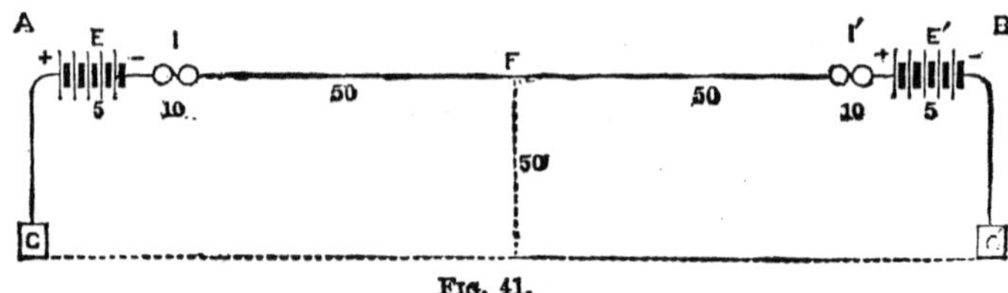

FIG. 41.

By Ohm's law we find the strength of this current to be as follows:

5	ohms resistance of battery, . .	= R
10	" " " instrument,	= L
50	" " " ½ line,	
50	" " " fault,	
115	= R + L.	

$$C = \frac{E}{R + L} = \frac{1000}{115} = 8.7$$

With the key *closed* at A, the strength of the current in the instrument at B was found to be

	15.4
With key open at A, as above........	8.7
Difference, or effective working force...	6.7

In this case the latter will obviously be the same, whether A sends to B or B to A.

104. SECOND CASE.—Suppose the same fault to be located near A (see Fig. 42).

The current from the battery E will divide at F, part going to the ground through the fault, and the remain-

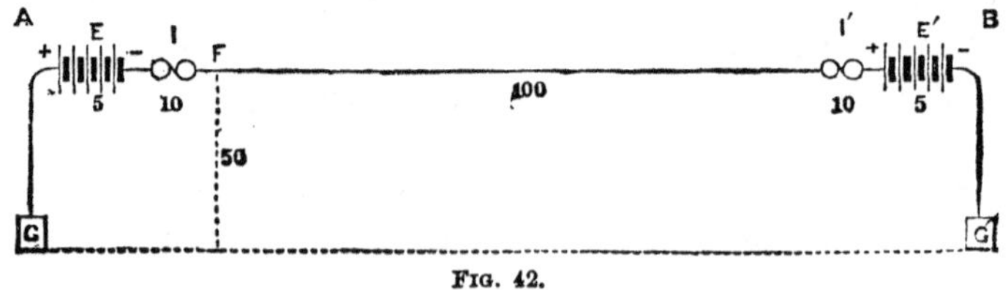

FIG. 42.

der over the line to B, and through the instrument and battery to ground. The current from E′ will divide in the same manner between the fault and the route through I and E. Taking the battery E alone, and considering the other battery E′ simply as a conductor, the two circuits beyond the fault give the following resistance :

1. By the line instrument and battery at B.. 115 ohms.
2. " fault F........................ 50 "

Their joint resistance will be * ..... $\frac{115 \times 50}{115 + 50} = 34.8$ ohms.

Add resistance of battery itself, 5 ohms, and instrument, I, 10 ohms .... 15 "

The total resistance will be........................ 49.8

And the current leaving the battery, E, $= \frac{1000}{49.8} = 20$

This current will divide at the fault between the two circuits, whose resistances are respectively 115 and 50, or in the proportion of 23 to 10. Therefore 23 parts of the current will go to the ground at F, and 10 parts, $= \frac{20 \times 10}{33} = 6.1$, will go over the line to B.

The current from the other battery, E′, in like manner divides at F, between the fault and the circuit through the instrument and battery at A. The joint resistance of the two circuits is

$$\frac{15 \times 50}{15 + 50} = 11.5$$

Add the resistance of the battery E, 5 ohms, instrument I, 10 ohms, and line, 100 ohms 115.0

Total resistance 126.5

The current leaving the battery E will therefore be $\frac{1000}{126.5} = 7.9$

The resistance of the two circuits beyond the fault being 15 and 50, or as 3 to 10, 3 parts will go to ground and 10 parts, or $\frac{7.9 \times 10}{13} = 6.1$, through I.

---

* The *joint resistance* of any two circuits is found by *dividing the product of the two resistances by their sum.* When there are three circuits, first find the joint resistance of two circuits as above, and treat it as a single circuit, again applying the same rule. In the same manner the joint resistance of any number of circuits may be calculated (175).

When A sends to B, the current in the instrument at B will be :

Key closed at A.		
From battery E′........................	7.9	
" " E........................	6.1	
Total strength in I′.....................		14.0
Key open at A.		
From battery E′............... $\frac{1000}{165}$ =	6.1	
" " E........................	0.0	6.1
Difference, or available working current at B,		7.9

Now let B send to A. The current at A will be :

Key closed at B.		
From battery E........................	20.0	
" " E′........................	6.1	
Total strength in I.....................		26.1
Key open at B.		
From battery E............... $\frac{1000}{65}$ =	15.4	
" " E′........................	0.0	
Total strength in I............		15.4
Difference, or available working current at A,		10.7

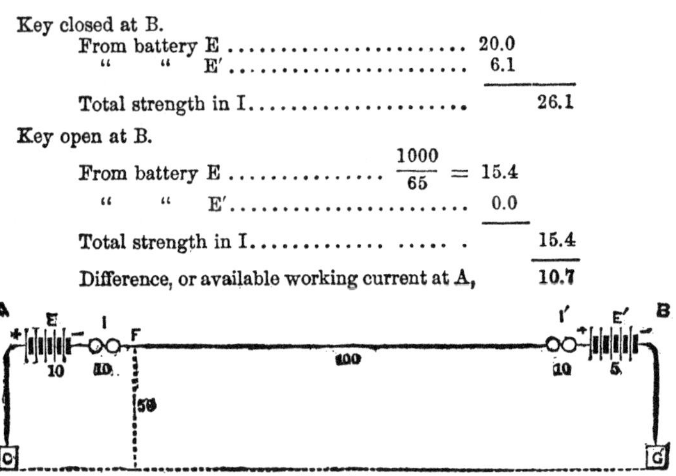

FIG. 43.

105. THIRD CASE.—Let the battery at A be doubled, the fault remaining as in the last case. The electromotive force and internal resistance of E are both doubled, as in Fig. 43. The current from E will now be :

$$\frac{2000}{20 + \frac{50 \times 115}{50 + 115}} = 36.5$$

which will divide at the fault in the same proportion as before, the part going to B being $\frac{36.5 \times 10}{33} = 11.0$.

The current from E′ will be $\frac{1000}{115 + \frac{20 \times 50}{20 + 50}} = 7.7$, and the portion reaching A $\frac{7.7 \times 5}{7} = 5.5$.

When A sends to B the signals will be as follows:

Key closed at A.

Current at B $= 7.7 + 11.0 = 18.7$

Key open at A.

Current at B...... $= \frac{1000}{165} = 6.1$

Effective strength at B......... 12.6

Now let B send to A:

Key closed at B.

Current at A $= 36.5 + 5.5 = 42.0$

Key open at B.

Current at A....... $= \frac{2000}{70} = 28.6$

Effective strength at A......... 13.4

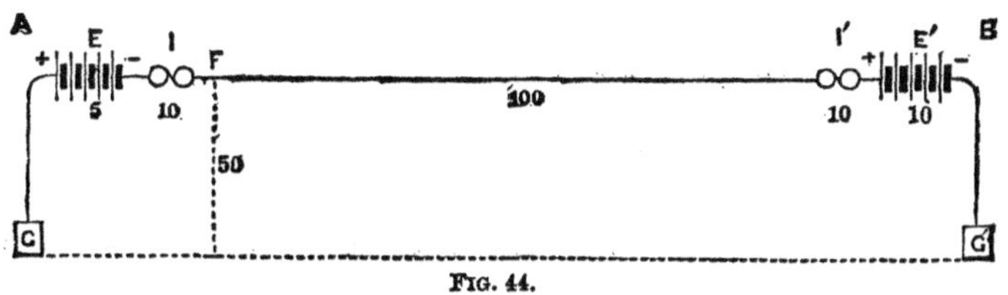

FIG. 44.

106. FOURTH CASE.—Double the battery at B, the fault remaining unchanged. See Fig. 44.

$$\text{Current from E} = \frac{1000}{15 + \frac{50 \times 120}{50 + 120}} = 19.9$$

$$\text{Portion going to B...} = \frac{19.9 \times 5}{17} = 5.8$$

$$\text{Current from E}' = \frac{2000}{120 + \frac{50 \times 15}{50 + 15}} = 15.2$$

$$\text{Portion going to A..} = \frac{15.2 \times 10}{13} = 11.7$$

A sending to B:

Key closed at A.

Current at B $= 15.2 + 5.8 = 21.0$

Key open at A.

Current at B........ $= \frac{2000}{170} = 11.8$

Effective strength at B.......... 9.2

B sending to A:

Key closed at B.
Current at A = 19.9 + 11.7 = 31.6

Key open at B.

Current at A........ = $\frac{1000}{65}$ = 15.4

Effective strength at A........... 16.2

107. Thus we find that on a circuit consisting of

Line wire resistance....................	100 ohms.
2 batteries "	10 "
2 instruments "	10 "

each battery having an electro-motive force of 1000, the signals received will be as follows:

	Signals at A.	Signals at B.
When the line is perfect....................	15.4	15.4
With escape 50 ohms in centre	6.7	6.7
Same fault at A...........................	10.7	7.9
Same fault at A, with battery doubled at A	13.4	12.6
Same fault at A, with battery doubled at B.....	16.2	9.2

108. The results of this investigation may be summed up as follows:

When the batteries and instruments are equal at each end of a line, a given fault will interfere most with the working of the circuit when in the centre.

When the fault is near one end of the line, the station farthest from it will receive the weakest signals, and the station nearest it the strongest signals.

In increasing the battery power for working over an escape, the addition should be made to the battery nearest the fault.

109. Distribution of Battery Power.—If the insulation of a line was perfect at all times, the position of the battery in the circuit would be a matter of indifference. As all lines, however, are subject to more or less leakage or escape throughout their entire length, the whole battery should not be located at one end of a long line, for in this case signals would be received much better at one end of the line than the other. The usual arrangement is to place half the battery at each

end of the line, although if the escape be uniform throughout the entire length of the line, the effect upon its working will be the same, whether all the battery is placed in the centre of the line or a portion of it in the centre and the remainder divided equally between the two ends.

If a certain portion of the line is especially defective in its insulation, the distribution of battery power may sometimes be varied in accordance with the principles laid down, with manifest advantage.

The insulation of the batteries themselves is a matter of great importance, and should never be neglected. (29.)

110. Working several Lines from One Battery.—It has been for many years the practice in this country to work a considerable number of lines at the same time from a single battery. The number of wires that can be worked in this manner without interference depends entirely upon the proportion between the internal resistance of the battery employed and the joint resistance of all the circuits connected with it. If the resistance of the battery itself is inappreciably small in comparison with that of the lines connected with it, the current on any given circuit will vary but little, whether the others be open or closed. With the Grove battery of, say, 50 cups, it is possible to work as many as 40 or 50 well insulated lines, of 300 miles or more in length, without appreciable interference. The great objection to this system is that, in wet weather, the resistance of the lines is enormously diminished, and the interference of one circuit with another, as a necessary consequence, greatly increased.

It is a common practice when this occurs to increase the *number* of cups in the battery, which in most cases has a tendency to aggravate the very evil it is sought to remedy; for with every such addition the resistance of the battery becomes greater in proportion to that of the lines, and the currents more unsteady and fluctuating. No small part of the trouble experienced in working lines in wet weather arises from this cause, although usually attributed entirely to defective insulation. It is true, however, that the latter indirectly causes the difficulty, by lessening the resistance of the wires.

111. Experiments made on a very wet day, upon a number of circuits of nearly the same length (100 miles), leading out of New York city, proved that when one such wire was attached to a carbon battery of 60 cups the addition of three other similar wires reduced the current on the first one 12 per cent. It is a common practice to attach as many as eight wires to such a battery, which in the above case would have reduced the current about 25 per cent.

112. It is the opinion of many scientific experts in practical telegraphy that increased efficiency, as well as economy, would result from working telegraph lines with a single series of Daniell's battery, in its most approved form, upon each circuit. The objection urged against this battery is the increased amount of room it takes up, as well as its somewhat greater original cost.

113. As long as the present system remains in vogue, care ought to be taken that the different circuits leading from the same battery are as nearly as possible equal in resistance; and it must not be forgotten that the interference caused by attaching too many wires to a battery *cannot be remedied by the addition of more cups for intensity*. The electro-motive force of a carbon battery is exhausted with a rapidity nearly in proportion to the number of circuits supplied from it. In the case of the Grove battery this effect is not so apparent.

CHAPTER VI.

TESTING TELEGRAPH LINES.

114. Interruption and interferences from various causes are constantly occurring upon telegraph lines, and one of the most important of an operator's duties is to be able to discover promptly the nature and location of a fault, that measures may immediately be taken for its removal. This is done by an investigation called *testing*. The apparatus and methods now in general use in this country are of a somewhat primitive nature, but the improved modes of testing which have long been employed in Europe are gradually becoming appreciated here, and as these are based on sound scientific principles, it is to be hoped that they will soon supersede the imperfect ones heretofore employed.

115. The principal interruptions to which a telegraphic circuit are liable may be summed up as follows:

DISCONNECTION.—The continuity of the circuit is broken, so that no current passes over the line.

PARTIAL DISCONNECTION.—This is usually caused by rusty and unsoldered joints in the line, or by loose screw connections in offices or about switches, which offer great *resistance* to the passage of the current.

ESCAPE or leakage of current from the line to the ground, caused by defective insulation, or contact with trees, &c. When an escape is sufficient to entirely prevent the working of the line it is called a "*ground.*"

CROSS.—When two wires are in contact, so that one cannot be worked without interfering with the other.

WEATHER CROSS.—When a portion of the current from one wire leaks into others upon the same poles, through defective insulation. The effect is similar to that of a cross, but much less strongly marked. This is often improperly called *sympathy*, or *induction.*

DEFECTIVE GROUND CONNECTION.—It sometimes happens that the ground wire, or ground plate, at a terminal station, is defective. The effect of this is to make the wires connected with it appear as if in contact, or *crossed.* This difficulty is often caused by the removal of a meter in offices where a gas pipe is used as a ground connection for several lines.

116. TESTING FOR DISCONNECTION.—If the circuit is broken at any point the relays will all remain open. The operator at each way station should immediately proceed to test the wire by connecting his ground wire, first on one side of the instruments and then on the other. If either connection closes the line circuit, the interruption is on that side, as the circuit of the opposite main battery is completed through the ground, in place of the broken wire. If the ground wire gives no circuit either way, it is probable that the interruption is in the office, or that the ground connection is defective. Each operator should always first make sure that the fault is not in or about his own office. Having ascertained the direction in which the difficulty lies, he should at once report the state of the case to the terminal station at the opposite end.

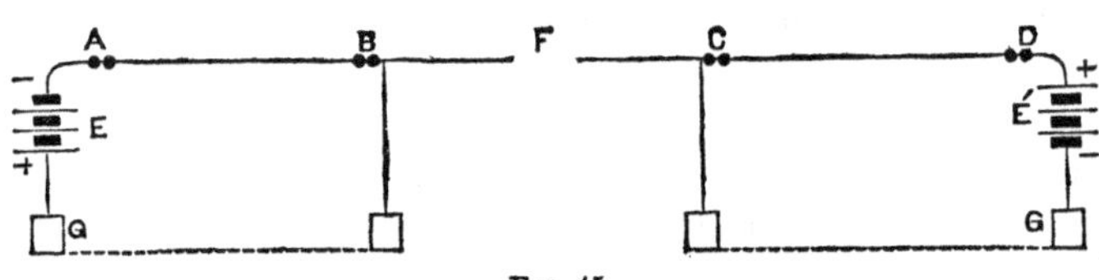

FIG. 45.

Fig. 45 represents a line with four stations, A, B, C and D. Suppose the wire broken at F. By connecting the ground wires at B and C, as shown, two distinct circuits are formed. A can work with B, and C with D, showing that the fault is between B and C.

Disconnection is usually caused by the breaking of the line wire, or by a key carelessly left open. Some other causes are wires loose in their binding screws, or defective switches, or the fine wire in or about the relay may be broken. The latter is sometimes burned in two by atmospheric electricity.

117. Partial Disconnection.—It is rather difficult to discover this fault by the ordinary relay tests. It is frequently of an intermittent character, and requires to be very carefully tested for. In the latter case, the best plan is to cross connect, or interchange the defective wire with a good one at the terminal, and also one other station, as in Fig. 46. Suppose the fault is at F, on No. 2 wire; by cross connecting at A and B, as shown,

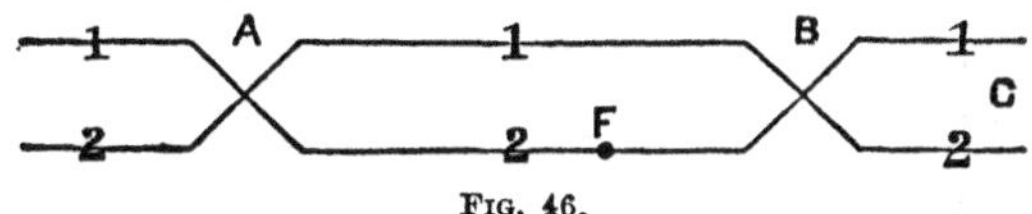

Fig. 46.

the fault will shift to No. 1 circuit, showing that it is between those points. If it were beyond B, it would remain on No. 2 circuit. In this case, let the wires be put straight at B, and cross connected at C, and so on, station by station. When the fault is passed it shifts to the other circuit, and will therefore be found between the two last stations.

118. To Test for an Escape.—Call the stations up in rotation, beginning with the one farthest off, and have them open key for a minute or two. When a station beyond the escape is open, more or less current will still pass out to the line through the relay, returning through the ground from the fault.

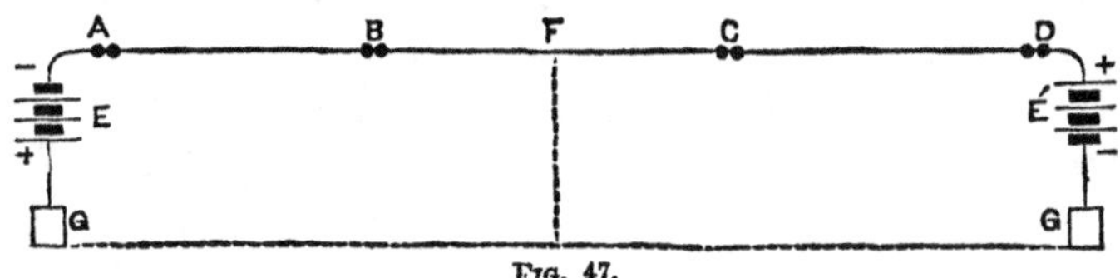

Fig. 47.

Suppose A (Fig. 47) is testing. When the circuit is open at C or D a current will pass from E through the fault, F, which will be interrupted when B opens, showing the fault is between B and C.

119. TESTING FOR GROUNDS.—A ground is tested for in a similar manner. The operator at a way station can ascertain which side of him the ground is situated, from the fact that it cuts off or greatly weakens the main battery current from that direction, when tested with a ground wire by means of the finger or tongue.

Telegraph lines are very liable to be grounded by the action of atmospheric electricity upon the lightning arresters. These should be carefully looked after at frequent intervals, especially where exposed to dampness, as in cable boxes.

120. TESTING FOR CROSSES.—In case a cross is suspected between two wires, say Nos. 1 and 2, instruct the most distant station to open one wire, preferably the through wire, or No. 1, and "send dots" upon the other. Open No. 2 at your own station, and if the dots sent on No. 2 at the distant station are received on No. 1, the wires are crossed. Care must, of course, be taken not to be deceived by the leakage from one wire to another, caused by defective insulation. If the wires are in actual contact, the signals received upon No. 2 wire will be nearly or quite as strong as if received upon No. 1.

Next, instruct the distant station to leave No. 1 open, and open it also at your own station. No. 2 will now be free from interference, and the stations upon it may be signalled without difficulty. Call them in regular succession, commencing at the farthest end oı the line, and instruct each one in turn to send dots on No. 2. If the dots are received on both wires the cross is between you and the station sending; but if upon No. 2 only, it is beyond that station. It is better that each operator, while sending dots, should open the other wire, if practicable.

The principle of this test will be understood by reference to Figs. 48 and 49, which represent a two wire line, with four stations, A, B, C and D, the wires being "crossed" between B and C. The operator testing for the cross is supposed to be at A. In Fig. 48 station C has No. 1 open, and station A has No. 2 open. If C sends dots on No. 2 the circuit will shift to No. 1, at the cross, as shown by the arrows, and the dots will

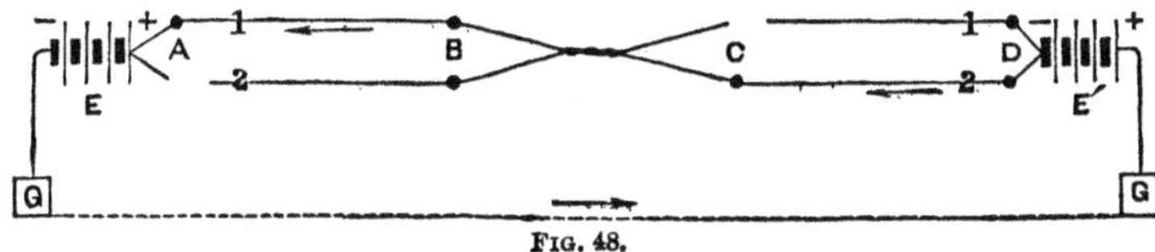

FIG. 48.

come on No. 1 instrument at A, showing that the cross is between A and C. In case C were unable to open No. 1 the effect would evidently be the same, provided it remains open at D.

Now let C close both wires, and B open No. 1, and write dots on 2 (Fig. 49). If No. 2 be open at A, B will be unable to work in this case, as both wires are open, one at A and the other at B. With both wires

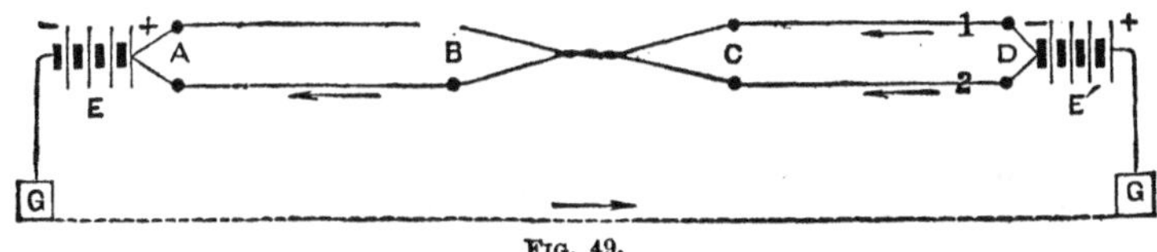

FIG. 49.

closed at A, B's dots will come on No. 2, the current from F passing over both wires to the cross, and from thence on No. 2, No. 1 being open, as shown in the figure. Thus the fault is located between B and C.

In large offices, where there are a considerable number of wires, it will often be found a much more convenient and expeditious method of testing for crosses for the operator to station himself at the switch with a single instrument, which can be placed at pleasure on any wire, for the purpose of communicating with different stations. When any station is "sending dots" the testing operator can feel them by placing a finger upon the ground wire, and another upon the proper line wire at the switch. The principle involved is of course the same as in the method first described. In wet weather, however, testing by the sense of feeling is attended with much uncertainty, as it is impossible to distinguish between the effect of a metallic cross, or actual contact, and the leakage arising from bad insulation.

121. It would be difficult to specify all the minor interruptions that are liable to occur in and about telegraph offices, or on the lines; the operator will therefore, in many cases, be obliged to depend upon his own ingenuity for the best method of testing applicable to each particular case. By carefully studying, however, the principles heretofore explained, the intelligent telegrapher will usually be able to cope with any difficulty that may chance to arise in the ordinary service of the lines.

122. Testing with the Galvanometer and Resistance Coils.—In the more accurate and scientific modes of testing, which have been for some years em-

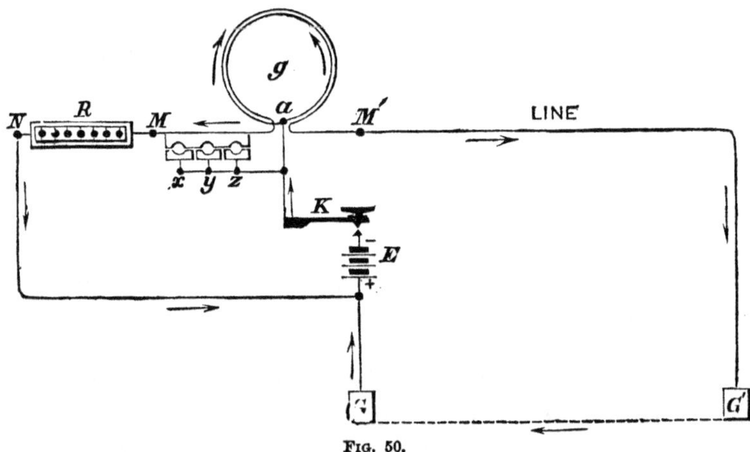

Fig. 50.

ployed upon the European lines, the instruments used are the *differential galvanometer* and a set of standard *resistance coils* (44).

The arrangement of the connections will be understood by reference to Fig. 50.

The galvanometer coils are wound with two wires of the same length and resistance, insulated from each other with the utmost care. The needle is, therefore, surrounded by an equal number of convolutions of each wire, which are also equi-distant from it.

The inner end of one coil surrounding the galvanometer, g, is joined to the outer end of the other at a, and a key, K, attached to this junction, when depressed, forms the connection with the testing battery, E. The other ends of the coils run to binding screws, M M′, for the convenient attachment of lines to be tested. The principle upon which the action of the instrument depends is the following:

When the battery is connected by depressing the key the current divides into two equal portions at a, one flowing from a to M, tending to deflect the needle to the left, and the other from a to M′, tending to deflect it to the right; but as long as the two currents are of the same strength they balance each other, and the needle remains at rest. Suppose that the terminal M′ is connected to a telegraph line whose remote end is to ground, and the terminal M is connected through the resistance coils, R, to the ground likewise, as shown in Fig. 50. We have already seen (102) that the strength of an electric current is in all cases equal to the electromotive force divided by the resistance. Therefore, if in this case we let

E = electro-motive force of battery,
l = resistance of line wire,
g = resistance of galvanometer coil,
r = resistance coils in circuit,

the current in the two circuits surrounding the needle will be

$$\frac{E}{g + l} \text{ and } \frac{E}{g + r}$$

If, therefore, the resistance coils in circuit be varied until $r = l$, the needle will remain unaffected. As the earth offers no appreciable resistance to the passage of the current, the resistance of the line l will be accurately represented by the amount of resistance interposed at r, in order to bring the needle to zero. The value of the above equation will obviously not be affected by any change in the value of E.

123. The resistance coils, R, accompanying the galvanometer, are so arranged as to be adjustable to any required resistance, from 1 ohm up to 10,000. For the measurement of still higher resistances one coil of the galvanometer is provided with three "shunts," or branch circuits, x, y, z, having resistances respectively equal to $\frac{g}{9}$, $\frac{g}{99}$, and $\frac{g}{999}$; therefore, if x be connected, $\frac{1}{10}$ of the current will pass through g, and $\frac{9}{10}$ through the shunt. In the same manner y and z respectively allow but $\frac{1}{100}$ and $\frac{1}{1000}$ of the current to pass through one wire of the galvanometer, when connected, and by this means the instrument may be made to measure any resistance, from 0.001 up to 10,000,000 ohms.

124. Testing for the Distance of Faults.—The principle upon which the methods of *distance testing* are founded is that of finding the resistance of the line wire between the testing station and the fault by means of the apparatus described. When the line is broken at any point one of the following four cases generally occurs:

1. Line wire broken, giving full, or nearly full, ground connection.
2. Line wire unbroken, but gives nearly enough escape to ground to make signals imperceptible.
3. Line wire broken, without making contact with earth.
4. A cross between two wires, so that signals sent on one are communicated to both.

125. It is very essential that the resistance of each circuit should be frequently measured and recorded, so that when a fault occurs the actual resistance per mile of the line may be known. If the broken line gives a full ground, its resistance divided by the resistance per mile at once gives the distance of the break from the testing station; and if the distant station obtains a

corresponding result, the confirmation is complete. Thus, in a line of 100 miles in length, if the tests from the two extremities indicate distances of 45 and 55 miles, respectively, the locality of the interruption is clearly indicated. As the fault, however, usually gives a very considerable resistance at the point where the line is in contact with the earth, and the sum of the two resistances, measured from stations at the opposite ends of the line, greatly exceeds the resistance of the line itself when perfect, it is usual in such cases to estimate the fault midway between the two points indicated. Thus, when the respective resistances indicate 86 and 26 miles, the sum of these exceeds 100 miles by 12, and therefore half this excess, or 6, is deducted from each of the measures.

126. When the line is unbroken, but shows a heavy escape or partial ground, sufficient to weaken signals, two or three different methods are available for determining its locality. The first plan is that of direct measurement, alternately from each end, the distant end at the same time being insulated, or, in other words, left open, in the manner explained in the last paragraph. In this case the resistance of the fault is measured twice over, and is roughly allowed for by the method of calculation above given.

127. The Loop Test.—A second and more accurate method, which gives a measure entirely independent of the resistance of the fault itself, is known as the loop test. It is only available, however, in cases where there are two or more parallel wires on the same route. In order to make this test, let the operator proceed as follows:

Make the length to be tested as short as possible, and have all the instruments in circuit taken out. Select a good wire, similar, if possible, to the one it is required to test. Both these wires must then be connected together in a loop, at the nearest available station beyond the fault, *without* ground connection. The resistance of the faulty wire, when perfect, must be ascertained. This may be taken from previous records, or it may be found thus:

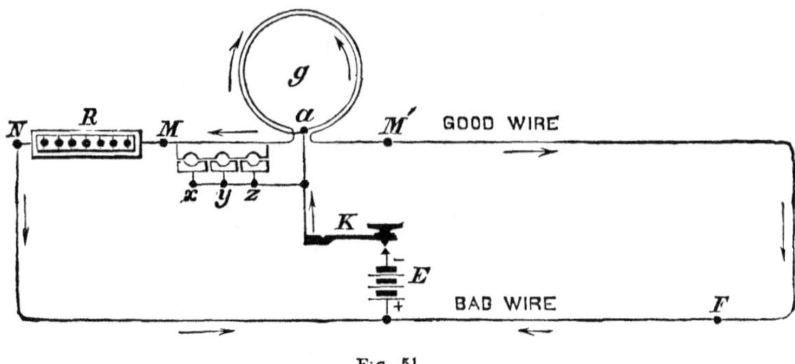

Fig. 51.

Connect one end of the loop to the + pole of the battery E, and the other to one of the galvanometer wires at M′. Connect also the + pole of the battery with the resistance coils, R, at N, and the – pole of the battery to the key, K, and common terminal, *a*, of the galvanometer. Connect the remaining galvanometer wire with the resistance coils at M (Fig. 51).

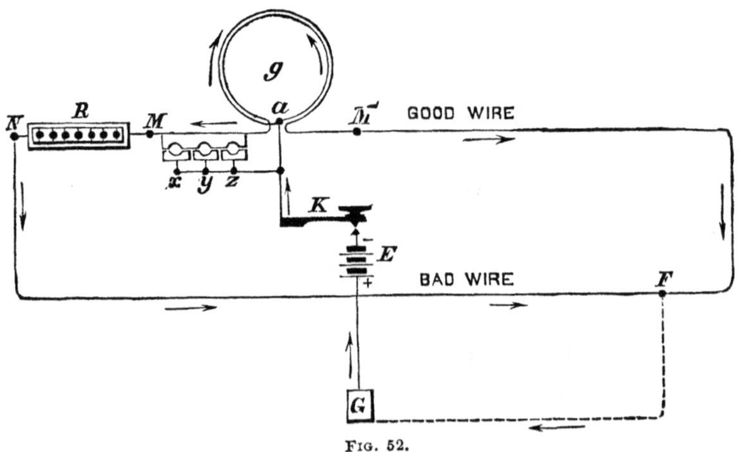

Fig. 52.

Having ascertained the resistance of the loop, arrange the connections as shown in Fig. 52.

Upon depressing the key, K, the battery current will

flow through M and R, and also through the loop. The resistance required at R to balance the needle will be equal to the sum of the resistance of the two lines. Although there is a partial ground at F it will not affect the measurement, as there is no other ground in circuit.

Connect the + pole of the battery E to ground, and connect the — pole to the key K. Connect the perfect wire of the loop to one of the galvanometer wires at M', and the faulty wire to the other galvanometer wire at M, interposing the resistance coils R. When the key K is depressed the current from the battery E flows into both lines simultaneously, passing to the ground through the fault at F. By adding resistance at R, so as to bring the needle to zero, the resistance *a* N F will be made equal to *a* M' F.

The resistance thus added, deducted from the total resistance of the loop, previously ascertained, and divided by two, is the resistance of the line between N and F.

Thus, if the resistance of the loop be 1,000 ohms, and 100 ohms have been added to the defective wire to balance the needle—

Then $\frac{1000 - 100}{2} = 450$ ohms, the resistance of the wire between the resistance coils R and the fault.

Let M' F $= x$, N F $= y$.
Then $x + y = L$, the resistance of the loop.
As $R + y = x$, or R + N F = M' F.

$$\text{Therefore, } y = \frac{L - R}{2}$$

Suppose that the loop of 1,000 ohms measures 120 miles, then, by proportion,

If 1,000 ohms = 120 miles, 450 ohms = 54 miles.

When an instrument or section of small wire is included in the circuit, allowance must be made for their resistance. It is a great assistance in these tests to know from previous records the exact resistance of every section of the line.

128. Blavier's Formula for Locating an Escape.—Where there is but one wire the following method may be employed. Three tests have to be taken for the operation, viz:

Let R = resistance of the line before it was defective. This must be obtained from previous records.
" S = resistance of the line when grounded at the distant end.
" T = resistance of the line when disconnected at the distant end.

Multiply S by S and T by R, and add the products together; subtract from this amount T times S, and also R times S. Subtract the square root of the remainder from S; the remainder will give the resistance, x, or the distance of the fault from the testing station.

This process appears complicated, but is in reality very simple. For example, suppose the line 100 units long, and the fault 68 units distant, and the resistance of the fault 96 units, as shown in Fig. 53

x 68 y 32 z 96

Fig. 53.

$$\begin{array}{llllllll} \text{Then R} & = & x + y & = & 68 + 32 & = & 100 \\ \text{*S} & = & x + \dfrac{y \times z}{y + z} & = & 68 + \dfrac{32 \times 96}{32 + 96} & = & 92 \\ \text{T} & = & x + z & = & 68 + 96 & = & 164 \end{array}$$

We shall, however, have obtained these resistances by measurement. and not by calculation. We therefore have:

$$\begin{array}{l} \left.\begin{array}{l} S \times S = 92 \times 92 = 8464 \\ T \times R = 164 \times 100 = 16400 \end{array}\right\} 24864 \\ \left.\begin{array}{l} T \times S = 164 \times 92 = 15088 \\ R \times S = 100 \times 92 = 9200 \end{array}\right\} \begin{array}{r} 24288 \\ \hline 576 \end{array} \end{array}$$

And the square root of 576 is 24; which deducted, S = 92, gives 68 as the resistance of x, or the distance of the fault from the testing station.

The distance, x, being known, the others are obtained with ease; for R — 68 gives y, the distance from the opposite end; and T — 68 gives z, or the resistance of the fault itself. This test should be taken from both

* See note, § 104.

ends of the line, if possible. In the above calculation the resistance of the fault is supposed to remain constant during the measurements; but as this is not often the case in practice, the average of several measurements should be taken.

129. To Find the Distance of a Cross.—The two wires in contact form a *loop*, provided they are clean, and are twisted together, so that the contact offers no appreciable resistance. In such a case open both wires at the nearest station beyond, and test the resistance of the loop. Half this resistance will be the resistance of the wire between the galvanometer and the fault, and from this the distance can be calculated, as before explained (127). All relays in circuit must be taken out, or the proper allowance made for their resistance.

As it is difficult to tell with certainty whether the cross offers resistance or not, it is a better plan to test it as a ground. Test each wire in turn by the loop method (127), grounding the wire at both ends. The wire tested will then make a ground through the other wire at the point of contact, and the location of the latter may be readily ascertained.

*Second Method.**—Suppose two wires, A and B, touch one another at the point F. Connect A to the zinc of the testing battery, leaving it open at the remote end; it will then serve as a battery wire between the battery and the fault (F). Ground B at the distant end, and connect it to one coil of the differential galvanometer at the testing station Put the other wire of the galvanometer to ground. The current of the battery will pass along the wire A and divide at F, one portion going to ground at the distant end of B, and reaching the galvanometer through the wire connected with the ground, the other portion returning to the galvanometer through the nearer portion of B. If the cross is exactly in the centre of B the needle will not move, as the two currents will balance each other. If one section of B is longer than the other, the resist-

* Culley's Handbook, 3d edition, page 279.

ance added to the shorter section to balance the needle will show the difference in the resistance of the two sections.

Let L = the total resistance of B.
" x = resistance of the shorter portion.
" L − x = " " longer "
' R = resistance added to shorter portion.

$$\text{Then } x = \frac{L - R}{2}$$

130. Advantages of Testing by Measurement.—The testing of lines by actual measurement lies at the very foundation of all efforts to improve the working of our telegraphic system. The insulation resistance of each of the principal circuits should be measured every morning, and a record of the results kept for reference. In England the standard of insulation is 1,000,000 ohms per mile in the worst of weather. Therefore, a line of 200 miles should not give less than $\frac{1,000,000}{200} = 5,000$ ohms. If it gives less than this the low resistance is due to defective insulation. The line should, in that case, be tested in many separate sections, either from the terminal office or by a visit to each section. If the resistance per mile is the same for each section, the fault is probably owing to the nature of the insulation; but if, as is usually the case, some sections are very much worse than others, the trouble will be found in contact with trees, broken insulators, and the like. A visit to the faulty locality will disclose the cause of the evil.

131. In comparing the insulation of line wires of different lengths, the insulation *per mile* must be ascertained, otherwise the longest wire will appear the worst; therefore, multiply the insulation test in ohms by the length of the wire in miles. If the insulation is uniformly good throughout the circuit tested, the leakage will increase in direct proportion to the length of the wire, irrespective of its thickness or conducting power, for the resistance of the wire is very small in comparison with the insulators, and need not be taken into account.

The following example from Culley's work will illustrate this. The figures given are the results of an actual test:

The wire A had a leakage equal to..............	29
" B " " "	30
" C " " "	50
Total leakage..........................	109

The three wires, when connected together at the testing end and left open at the distant end, gave a combined leakage of 110.

When connected so as to form a continuous wire, open at the distant end, the leakage was still 110. The experiment was repeated, and extended to other wires, with the same result. In this case the resistance of the insulators was very great compared with that of the wire—as much as two million ohms per mile. But on a wet day three similar wires, whose respective leakages were 196, 185 and 141, making a total of 552, when looped in a continuous line, as in the second case above, gave a test of only 476, the distant portion of the wire being in reality tested by a current weakened by the leakage in the nearer portions.

132. Testing for Conductivity Resistance.—The metallic resistance of the line wires should be occasionally tested in sections, in the finest weather. The resistance should be uniformly in proportion to the length of the wire. If any section discloses an unusually high resistance per mile, it is very probable that there are rusty, unsoldered joints in the line, or that the ground connections are defective. It is difficult for those who have not tried it to believe the vast improvement that may be made in any line in a few days by actual measurement, and an inspection of the sections which give indications of being defective.

It is not an uncommon occurrence to find that a single unsoldered joint in galvanized iron wire, which appears perfectly firm and sound, will give a resistance, when tested by the galvanometer, equal to many miles of line. A line containing many bad joints will frequently work better in wet than in dry weather, as the moisture increases the conductivity of the oxide between the wires at the joints.

In testing for conductivity, with the distant end of the wire to ground, as in Fig. 54, the result is sometimes interfered with by *earth currents.* It is therefore better, when practicable, to use the loop method, by connecting the wire to be measured in a loop with another wire of known resistance. Unless this test is made in fine weather, however, the leakage from one wire to the other will decrease the resistance of the loop. The battery must also be insulated from the

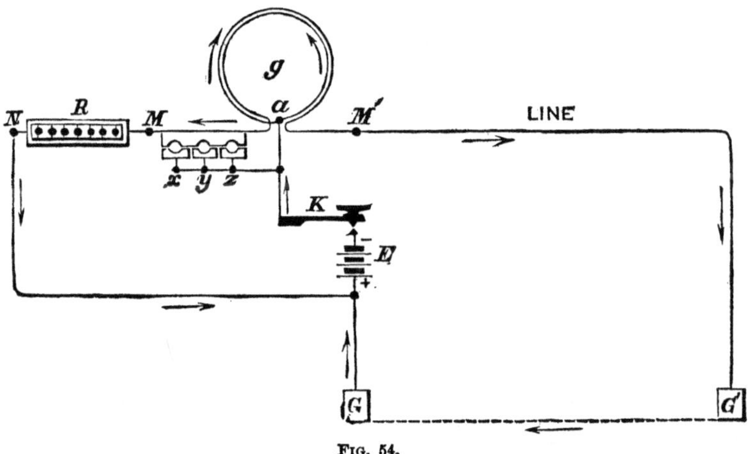

Fig. 54.

earth, otherwise the leakage at each insulator will decrease the apparent resistance, especially if the insulation is defective. For instance, two wires on the same poles, disconnected and looped at the distant end, had a resistance of 6,475 ohms when the battery was entirely disconnected from the earth. Upon putting the zinc pole of the battery and the line attached to it to earth, the apparent resistance fell to 5,250 ohms. The insulation resistance, with one wire disconnected, was 9,250 ohms, the weather being damp.

CHAPTER VII.

NOTES ON TELEGRAPHIC CONSTRUCTION.

133. In order to maintain uninterrupted telegraphic communication between any two points, it is of the first importance that the line should be well constructed and properly insulated throughout. There are numerous minor details in the construction and repairing of telegraph lines which merit much more attention than they generally receive. The bad working of our lines is in a great measure owing to the neglect of these apparently trifling details, through ignorance or carelessness.

134. Poles.—The poles intended for an ordinary line should never be less than five inches in diameter at the top, their length depending upon the number of wires to be provided for, and in some measure upon the location of the line. They should be set in the ground to the depth of five feet, wherever practicable.

In setting poles around the curve of a railway, they should be made to lean back against the strain of the curve.

135. Wire.—For ordinary lines, galvanized iron wire, of No. 8 or 9, Birmingham gauge, is generally employed. For short lines, No. 10 or 11 will answer very well. The "American Compound Wire," a recent invention, is composed of a combination of a steel core with a sheathing of copper. It has come into extensive use within the short time which has elapsed since its introduction, and has, thus far, been found to answer admirably. A wire of this kind, having a conductivity equal to a No. 8 iron wire, weighs but 112 pounds per mile.

136. The less the size, and consequently the conductivity of the line wire, the more care is required in its

insulation, for an increased resistance virtually adds to the length of the circuit. Increased conductivity thus admits of a reduction in battery power, with a consequent decrease in the escape of electricity, and long circuits may be thus worked with much greater facility; a fact which has been most unaccountably ignored in the construction of the greater portion of the lines in this country.

137. Galvanized or Zinc Coated Wire must always be used for permanent work, for rust reduces the conducting power of wires very rapidly. This is especially the case with the smaller sizes, such as No. 11 or 12. In smoky places it is a good plan to paint the wire before it is put up, for the gas arising from the combustion of coal destroys the zinc coating in a short time, as may be observed in many of our larger cities.

138. Arrangement of Wires upon the Pole.—Wires arranged vertically upon the poles, or one above another, are more liable to get into contact with each other than when arranged horizontally upon cross-arms. When placed one above another, each alternate wire should be fastened upon opposite sides of the poles.

It is better not to place wires of different sizes upon the same poles or cross-arms if it can be avoided, as they are much more likely to get "crossed" than wires of the same size would be, as they do not keep time with each other when swung to and fro by the wind.

139. Joints or Splices.—In the construction of a line nothing is of greater importance than the perfect continuity of the circuit, and this depends, in a great measure, upon the perfection of the joints. The importance of this has been very generally overlooked by the telegraphers of this country, and much trouble in working lines has been experienced in consequence, the cause of which has remained unsuspected. A single rusty unsoldered joint will often cause more resistance than fifty miles of line.

No joint or splice, however clean and firm, can be depended upon if made by mere contact or twisting.

Sooner or later the metals will certainly rust, and this tendency is increased by the passage of the current. When copper and iron wires are joined together the joint is especially liable to become defective from this cause. It is a common error to suppose that joints made in galvanized wire do not require soldering.

140. In making a joint each wire should be twisted round the other, in the manner represented in Fig. 55, the turns passing as close, and as nearly at right angles as possible to the wire which they surround. A wire must *never* be spliced by being bent back and twisted around itself.

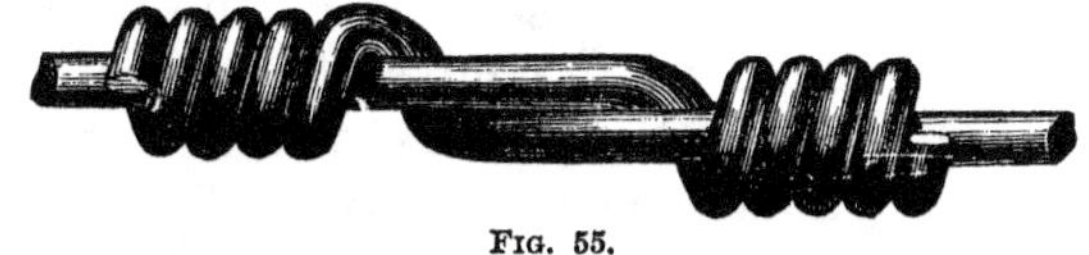

Fig. 55.

141. The best solution for soldering is chloride of zinc, with a little muriatic acid added, for the purpose of cleansing the wire. In connecting copper and iron wire together, it is well to wash off the chloride of zinc, and then coat the joint with paint or rosin, or else to solder with the rosin alone. This will prevent local galvanic action between the metals.

142. Fixing the Insulators.—In attaching insulators to the poles they should be arranged in such a manner as to prevent, as far as possible, the lodgment of snow about them, so as to form an escape between the wire and its support. The glass insulator is usually cemented to the bracket by means of white lead or asphaltum. *The edge of the insulator must never be permitted to touch the shoulder of the bracket;* for in this case, during a shower, a continuous stream of water flows directly from the wire to the pole, entirely destroying the usefulness of the insulator. For the same reason an insulator ought never to be fastened down to a bracket by means of a spike driven over it, as is often done where there is an upward strain upon the wire. The proper way, in such cases, is to use some form of

hook, or suspension insulator, and fasten the line into it with a tie-wire.

In turning a sharp angle it is better to put on two insulators and brackets at the corner pole, or the wire will be liable to come in contact with it.

When the Lefferts or Brooks insulator is used, there is danger of fracturing the glass while stringing wire, by violently wrenching the wire into the hooks. By a little precaution this result may be avoided.

143. Insulators and brackets are sometimes attached to a cross-arm, or other support, in a horizontal position. This ought never to be allowed, for a driving rain will wet the whole inner surface of the insulator, causing a great leakage of the current at every support. The same thing often occurs with improperly shaped brackets, which cause the spray from falling rain-drops to be dashed against the inside of the insulator. The shoulder of the bracket ought to be rounded or sloped off, so as to prevent this from happening.

Unless the insulator is securely fastened to the pin or bracket which supports it, it is liable to be lifted off by the wind, causing an interruption.

144. Leading Wires into Offices.—The wires leading into offices are fruitful sources of escapes and other interruptions, as the work is often very unskilfully or carelessly done. Gutta-percha covered wires, unless well protected, become entirely useless in a year or two, if exposed to the air and light. The method employed in England to protect this kind of wire might be adopted with great advantage in this country. The gutta-percha wire is first covered with tape, and then saturated with a preservative mixture.*

The best way to lead wires through the side of a building is to enclose them in hard rubber tubes, with

* This mixture is made and applied as follows: Take equal portions of wood tar, gas tar and slacked lime. Boil these together, stirring them well while boiling, until the moisture is entirely driven out, which may be known by the subsidence of the frothing. When cool apply to the taped wire, and then cover the latter with dry sand. Hang the wire up to dry in the air, and in three or four days it will be ready for use. This coating resists sun and moisture, and effectually protects the gutta-percha.

the outer ends inclined downwards, to prevent moisture from entering. In arranging these wires, it should be borne in mind that the current will follow moisture and dampness along the outer surface of covered wire, unless it is so placed that the line of leakage is broken at some point.

145. Fitting up Offices.—In running wires inside an office, it is better never to allow two wires to touch each other, even when covered with an insulating coating, as this may be burned by lightning or otherwise rendered imperfect causing a cross-connection. The proper mode of arranging the office connections and running the wires to the instruments is shown in Figs. 17 and 18, pages 34 and 35. Splices in the office wires should be avoided as far as possible, but when required, they should be made by turning each wire eight or ten times around the other. A less number of turns answers for the line wire, because the strain tends to keep the joint pressed together. Great care must be observed in making the joint between the iron and copper wire, which must in *all cases* be soldered.

146. Ground Connections.—It is of the utmost importance that the ground plate at each end of the line should make a perfect connection with the earth. The plate must be large, and buried deep in wet soil below the reach of frost. A water or gas pipe makes an excellent ground connection. The ground wire should be attached outside the metre, as the latter is liable to be occasionally disconnected for repairs. It is advisable, whenever practicable, to form a connection both with the gas and water pipes. The connection should be carefully made and always well soldered.

147. Cables.—The shore ends of cables should be bedded well out to low water mark. Dig the trench to a good depth, and cover the cable with a piece of heavy plank or joist, and secure it well with heavy stones, laid at short intervals. If the covering be merely of sand it will soon wash away and leave the cable uncovered. *Never allow any portion of a cable to*

be exposed to sun or air, but cover it all the way from the box where the connection is made with the air line.

Cable boxes always ought to be made double (one box within another), in order to prevent wet from entering. The unskilful manner in which these are often arranged is a fruitful source of trouble in working lines.

Lightning arresters should be kept attached to cables *all the year round.* It is not uncommon for heavy lightning to occur in midwinter in this country.

148. MAKING JOINTS IN CABLES.—In splicing cables, or other gutta-percha wire, the following is the method recommended by the Bishop Gutta-Percha Company, who have manufactured the greater portion of the submarine cables in use in this country:

"Use gutta-percha one sixteenth of an inch thick, cut in pieces to suit the joint. Soften it in hot water, and keep it flat. Wipe the surface with a cloth. Heat the surface by holding it near a flat file or other iron, about as hot as a laundress's iron; if the iron causes the gutta-percha to smoke, it is too hot. When dry and a *little sticky*, wind two or three coatings of gutta-percha around the joint, taking care that each coating is perfect and each layer is dry; then smooth off and lap the joint well over on the gutta-percha on each side of the joining. *Use no spirit lamp, nor anything with a blaze.* When gutta-percha is burned it cannot be restored. Hot water joints are worthless. They will not stand, and will open when dried out.

"In making joints it is *absolutely necessary* that the hands of the operator should be clean, and that no water, grease, dirt, or anything of the sort must be allowed to touch the gutta-percha."

149. Another method is given in *Culley's Handbook of Practical Telegraphy*, as follows:

"Prior to making the joint the gutta-percha is removed from the ends of the wires for about one and a half inches, and the copper wires are carefully cleaned by scraping; the wires are twisted together for one inch, the sharp ends being closely trimmed off. The joint is then soldered with *rosin* and good soft solder, containing a sufficiency of tin.

"After this the gutta-percha is scraped, or very carefully pared back for about two inches, to remove its outer surface, which is oxidized, and will not join properly; the wire joint is covered with Chatterton's compound* and the gutta-percha, heated on both sides, and tapered down over the joint till that from each side meets. The junction is completed by means of a warm joining tool, care being taken to mix the gutta-percha well without burning. As soon as this has cooled another coating of Chatterton's compound is spread over the gutta-percha, taking care not to burn the compound.

"A new and clean sheet of gutta-percha is then heated by means of a spirit lamp, and while so heated carefully stretched so as slightly to thin it. Then, while it and the Chatterton coated joint are still hot, it is laid on the joint, and pinched tightly round it with the finger and thumb, after which it is trimmed off close with scissors. The seam is again pinched and carefully finished off with a warm tool, so as to mix the gutta-percha of the two sides, and the coating of the wire itself, well together.

"The joint, when cool, is again covered with Chatterton's compound, and a longer and larger sheet of gutta-percha is laid over it, pinched, cut, and tooled off as before.

"When the joint is complete, another coating of Chatterton's compound is applied over the whole, well tooled over the joint, and when cool, rubbed with the hand, well moistened, till the surface is smooth.

"The mixing of the old and the new gutta-percha is most important, and joints generally fail from this having been imperfectly done, or from the percha being overheated. Cleanliness is essential to success. The fingers should be used as little as possible, and must be kept very clean."

* The ingredients of this are by weight, as follows: one part of Stockholm tar; one part of rosin, and three parts of gutta-percha.

CHAPTER VIII.

HINTS TO LEARNERS.

150. FORMATION OF THE MORSE ALPHABET.—The characters of the American Morse Alphabet are formed of three simple elementary signals, called the *dot*, the *short dash* and the *long dash*, separated by variable intervals or *spaces*. There are four spaces employed in this alphabet, viz., the space ordinarily used to separate the elements of a letter; the space employed in what are termed the "spaced letters," which will be hereafter referred to; the space separating the letters of a word; and lastly, that separating the words themselves.

The value of these spaces should be carefully impressed upon the mind of the learner. Beginners are apt to conceive that the Morse alphabet consists solely of dots and dashes, and this misconception has a tendency to greatly increase the time required to become good "senders." Uniformity and accuracy in spacing is of no less importance than in the formation of the letters themselves. The foundation of perfect Morse sending lies in the accurate division of time into multiples of some arbitrary unit.

151. The duration of a dot is the unit of length in this alphabet.

1. The short dash is equal to *three dots*.
2. The long dash is equal to *six dots*.
3. The ordinary space between the elements of a letter is equal to *one dot*.
4. The space employed in the "spaced letters" is equal to *two dots*.
5. The space between the letters of a word is equal to *three dots*.
6. The space between two words is equal to *six dots*.

The dot is an unfortunate appellation for this sign, because it conveys the idea of a point, or to speak electrically, a current of infinitely short duration. Electro-magnets, however, require time in magnetization (38). Currents involve time in transmitting signals. Clock-work requires time to run. Currents must be of sensible duration. The dot, therefore, involves *time*, but this time is variable, according to circumstances. The length of the dot should increase with the length of the circuit. In long submarine lines the dot has to be made longer than the dash itself on short open air lines, and the same thing occurs in working through repeaters (76). In commencing, therefore, the habit should be acquired of making short, firm *dashes*, instead of light, quick *dots*. After the student has once learned to send *well*, it is very easy to learn to send *fast*, but after once getting in the habit of sending short and rapid dots, or "clipping," it is almost impossible to get in the way of sending firmly and steadily. Beginners should rather take pride in the accuracy with which they space out the elements of the telegraphic music than in the number of words they can stumble through in a minute.

152. In the excellent little Manual of Prof. Smith* six elementary principles are laid down as the basis for practicing the alphabet, viz:

First principle. Dots close together.

I	S	H	P	6
..	...	....	.....	......

Second principle. Dashes close together.

M	5	¶
— —	— — —	— — — —

Third principle. Lone dots.

E
.

Fourth principle. Lone dashes.

T	L or cipher
—	——

---

* Published by L. G. Tillotson & Co., New York

Fifth principle. A dot followed by a dash.
A
. —

Sixth principle. A dash followed by a dot.
N
— .

153. Correctness in sending depends in a great measure upon the manner in which the key itself is handled. Place the first two fingers upon the top of the button of the key, with the thumb partly beneath it, the wrist being entirely free from the table. The motion should be made by the hand and wrist, the thumb and fingers being employed merely to grasp the key. The motion, both up and down, must be free but firm. *Tapping* upon the key must be strenuously avoided.

154. The downward movement of the key produces *dots* and *dashes;* the upward movement *spaces.* It is first necessary to acquire the habit of making dots with regularity and precision, then dashes, and finally combinations of dots and dashes. It is the best plan for the student to practice upon a register in a local circuit with his key, as he will the more readily be able to observe and correct the faults in his manipulation.

155. The student may now proceed to practice upon the elementary principles.

1. Practice making *dots* at regular intervals, until they are produced with the regularity of clock-work, and of definite and uniform dimensions. The regular tick of a watch or of a short pendulum is a valuable auxiliary in acquiring this habit.

2. Next proceed to make dashes, first at the rate of about one per second of time, which may afterwards be slowly increased to three. The space between the dashes must be made as short as possible. If the upward motion of the hand, in forming the space, be made *full*, it cannot be made too quick.

3. The third principle occurs but once in the alphabet, and forms the letter E. It is made by a quick but firm downward movement of the key. In practicing

upon this or any other character, it should not be repeated too rapidly, nor should the thumb and fingers be taken from the key in the intervals between the successive repetitions of the letter.

4. The fourth principle is somewhat difficult. The usual tendency is to make T too long and L too short. It will be observed that the same character is used for L and the cipher or 0. Occurring by itself or among letters it is always translated as L, but when found among figures becomes 0. This would at first seem liable to cause confusion, but in practice it is found not to be the case. It was formerly the custom to make the cipher equal to three short dashes.

5. The fifth principle, which forms the letter A, may be timed by the pronunciation of the word *again*, strongly accenting the second syllable. The tendency of beginners is usually to make the dot too long and the dash too short, and more especially to separate them too much.

6. The final principle, the dash followed by a dot, usually presents some difficulties. The universal tendency of the student is to separate the dot from the dash by too great a space. Time the movement by pronouncing the word *ninety*, with the first syllable somewhat longer than usual.

156. Having become thoroughly conversant with the six elementary principles, the following exercises may be taken up in order.

(1.)	E	I	S	H	P	6
	·	· ·	· · ·	· · · ·	· · · · ·	· · · · · ·

These should be practiced separately, until the right number of dots can be made invariably, the last dot in each being neither shorter nor longer than the preceding ones.

(2.)	T	M	5	¶	L or cipher.
	–	– –	– – –	– – – –	—

In practicing this exercise, care must be taken not to separate the dashes too much, and to make the final one in each letter exactly equal to the preceding ones.

Observe not to make the L too short. There is a general tendency in beginners to shorten the final dash, where two or more occur together.

(3.) A U V 4

The usual tendency to make too much space between the dot and dash, in the above letters, may be avoided by making them as if by prolonging the final dot in I, S, H and P.

(4.) I A S U
H V P 4

These are to be practiced in couples, as represented, the object being to impress upon the student the difference in the characters thus coupled together.

(5.) N D B 8

The student having thoroughly mastered the sixth elementary principle, he will have no difficulty in forming the above characters.

(6.) A F X Parenthesis
Comma Semicolon W 1

The only caution necessary in this exercise is to form the letters compactly, with the dashes of equal length. (See Exercise 2.) Observe, that the Parenthesis may be formed by running A U together, and the Semicolon by A F, etc.

(7.) U Q 2 Period 3

These differ but little from exercises previously practiced, and require no particular directions.

(8.) K J 9 Interrogation
G 7 Exclamation

J and K are generally considered the most difficult letters in the alphabet. Do not separate J into double

N, and be careful that the dashes correspond in length. (See Exercise 2.) The figures 7 and 9 require care in spacing correctly.

(9.) O	R	&	C	Z	Y
. .	. ..	. ...	.. .	... .	.. ..

These are termed the "spaced letters," and require great care in order to make them correctly. The "space" should be just double that ordinarily used between the elements of a letter. The usual tendency is to make it too great. It should be just sufficient to distinguish these characters from I, S and H.

157. The construction and manipulation of the alphabet having been thoroughly mastered by the practice of the foregoing exercises, it is now presented in its complete and consecutive form.

I. Alphabet.

A	.—	O	. .
B	—...	P	.....
C	.. .	Q	..—.
D	—..	R	. ..
E	.	S	...
F	.—.	T	—
G	——.	U	..—
H	....	V	...—
I	..	W	.——
J	—.—.	X	.—..
K	—.—	Y	.. ..
L	———	Z	... .
M	——	&	. ...
N	—.		

II. Numerals.

1	.——.	6	......
2	..—..	7	——..
3	...—.	8	—....
4	....—	9	—..—
5	———	0	———

III. Punctuation, etc.

*Period	··——··	Exclamation	———·
Comma	·—·—	†Parenthesis	·—··—
Semicolon	·—·—·	Italics	—···—
Interrogation	—··—·	‡Paragraph	————

Numbers are always sent twice over, to avoid error; once written out in full, and then in figures. In fractions one dot is used to represent the line between the numerator and denominator.

158. It is necessary to again caution the student against falling into the common error—from which most books on the telegraph are not exempt—that is entertained respecting the elementary signs of the Morse alphabet. It is said to consist of two characters, the dot and the dash. The importance of the *space* is utterly ignored. The difference between good and bad sending is almost entirely a matter of *spacing*. A common fault of young operators is to run their words too closely together.

If the principles laid down in this work be firmly adhered to, the learner will be surprised, not only at the rapidity with which he masters what appears to be a very difficult lesson, but at the extreme accuracy with which he manipulates his instrument. He must also carefully bear in mind that one of the most universal faults, among those attempting to learn the telegraphic art, is that of going over a great deal and learning nothing well.

159. Reading by Sound.—This can only be attained by constant and persevering practice, keeping in mind the principles above given. The lever of the Morse apparatus makes a sound at each movement, the down-

* The Semicolon, Parenthesis and Italics are seldom used in this country. It is customary among operators to emphasize particular words by separating the letters more widely than ordinarily.

† Preceding and following the words to which they refer.

‡ When this occurs the copyist makes a new paragraph, by commencing the next word upon another line.

ward motion producing the heavier one, or that representing dots and dashes; or, more properly, the heavy stroke indicates the commencement of a dot or dash and the lighter one its cessation. A dot makes as much noise as a dash, the only difference being in the length of time between the two sounds. Thus, if the recoil or lighter stroke be dispensed with, it would be impossible to distinguish E, T and L from each other.

In learning to read by sound it is best for two persons to practice together, taking turns at reading or writing, and each correcting the faults of the other. The characters must first be learned separately, and then short words chosen and written very distinctly and well spaced, the speed of manipulation being gradually increased as the student becomes more proficient in reading. After becoming sufficiently well versed in the art to read at the rate of twenty-five or thirty words per minute, the best practice will be found in copying with a pen and ink from an instrument connected with a line employed in transmitting regular commercial messages, in order that the student may familiarize himself with the usages of the lines and the minute details of actual telegraphic business.

In conclusion, the student is warned against falling into the common error of expecting great results from little labor. To become an expert operator requires much time and patience, and the most unwearied application. Remember, that whatever is worth doing at all is worth doing well. The time will seldom or never be found when a thoroughly competent operator cannot obtain immediate and remunerative employment, however overcrowded the lower walks of the profession may have become.

CHAPTER IX.

RECENT IMPROVEMENTS IN TELEGRAPHIC PRACTICE.

160. THE AMERICAN COMPOUND WIRE.—This important improvement in telegraphic conductors, referred to in another part of this work (135), has, within two or three years of its first introduction, become so extensively used that it seems likely in time to work a complete revolution in the American system of line construction. This wire is composed of a core of steel enveloped in a sheathing of pure copper, and coated with an alloy, of which tin is the principal ingredient, which serves to protect the whole from oxidization.

The relative strength of this wire is more than 50 per cent. greater than that of iron wire of equal weight, and its conductivity is also largely in excess of the latter. If we take, for example, a No. 8 galvanized iron wire, the gauge now usually employed in this country in the construction of the best lines, and compare it with a compound wire of nearly similar electrical capacity, the superiority of the latter will be manifest.

——	Weight per mile.	Tensile Strength.	Conductivity.	Poles per mile.
Galvanized Iron Wire (No. 8).....	375	1091	1	35
American Compound Wire (No. 8)..	112	514	1.07	23

In the above table the average conductivity of a mile of No. 8 galvanized wire is taken as 1, as a standard of comparison. The last column shows the number of poles per mile which will give the same percentage of strain upon the ultimate strength of the wire. In practice, however, it is safe to reduce the proportionate number of poles used for the compound wire, as the steel core is much more homogeneous and less liable to fracture on account of flaws, than the iron wire.

The advantages which arise from increased conductivity of the line wire and the diminution of the number of points of insulation and support are fully treated upon in another part of this work. The mechanical advantages of the compound wire are also very great. The labor of handling and stringing a light wire is much less than when a heavy one is employed. In running the wires over buildings, a mode of construction which has become very common in all large cities, stretches may safely be made double the length of those taken with the ordinary wire, and yet with less strain upon the insulators. Another important point in favor of this wire is the imperishable nature of the copper, which is the exposed metal. It is well known that the zinc coating of galvanized iron wire is soon destroyed near the sea-coast, and from the effects of carbonic acid arising from the combustion of coal in cities (137). Copper, under the same conditions, remains wholly unimpaired. Many cases occur in the construction of lines in which transportation is an item of great expense. In such cases, wire of the same or greater conductivity than galvanized iron, weighs materially less, with no disadvantage whatever arising from its lightness.

161. The following table exhibits the weight, size, and relative strength of compound wires, equivalent in conducting power to the ordinary sizes of iron wire used in telegraphic construction.

Size.	Galvanized Iron Wire.		Compound Steel and Copper Wire.			
	Weight per mile.	Relative Strength.	Weight per mile.	Relative Strength.	Size of Steel Core.	Size of Compound.
9	313	2.9	99	4.9	16	14
8	375	2.9	112	4.6	16	14+
7	449	2.9	121	4.4	16	13−
6	525	2.9	147	4.5	15	12−

The term relative strength, used in the preceding table, is the quotient obtained by dividing the strain which would break the wire by its weight per mile.

In constructing lines with the compound wire, much care should be used in making the joints so as not to separate the copper sheathing from the steel core, thus allowing moisture to penetrate to the steel and oxidize it. This may, however, be guarded against by carefully soldering the joints.

162. THE GRAVITY BATTERY.—Several modifications of the Daniell battery (19), especially adapted to telegraphic use, are finding much favor within the past few years. The most economical and generally useful of these improved forms is the gravity battery. The best arrangement is that known as the Callaud. Another combination very closely resembling it, and giving nearly as favorable results, is known in this country as the Hill battery. In these elements the porous cup of the Daniell battery is entirely dispensed with, the two

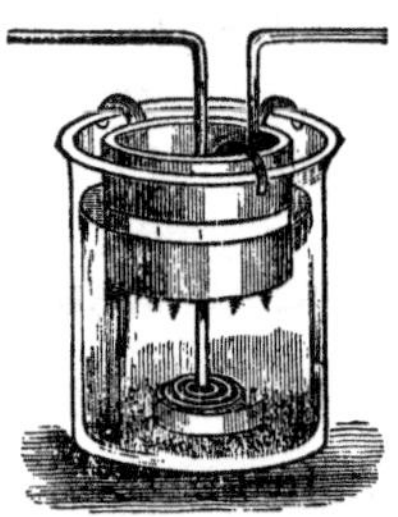

FIG. 56.

solutions being prevented from mingling by the difference of their respective specific gravities. The zinc plate of the Callaud element, in the form of a short hollow cylinder, open at both ends, is suspended in the upper portion of the containing jar, as shown in Fig. 56, by means of three hooks projecting from its upper edge, resting upon the jar. A strip of copper rolled into a spiral form is soldered to a copper wire covered with gutta-percha, forming the positive pole and connecting it to the zinc of the next element.

163. The manner of setting up this battery is as follows:

A sufficient quantity of soft water is poured into each jar to fill it to a point above the upper surface of the zinc. The battery should now be placed in the position which it is to permanently occupy, unless this has been already done. After the connections are made and everything in readiness, about three-quarters of a pound of sulphate of copper in lumps of the size of a hickory nut or larger, is dropped in, taking care that it does not lodge upon the zinc. The solution of sulphate of copper being of greater specific gravity, will remain at the bottom of the jar. The battery, after it is set up, should be kept on a closed circuit for about twelve hours, when its resistance will have become reduced so that the force will be available. As the battery continues in action, the sulphate of copper solution gradually becomes weaker and the zinc solution stronger. It is therefore necessary from time to time to add crystals of sulphate of copper, and to remove a portion of the zinc solution and replace by water. A good practical rule for maintaining this battery is to always see that the stratum of liquid around and in contact with the copper is kept of a blue color. The formation of transparent crystals upon the zinc indicates that the point of saturation of the zinc solution has been reached and that it should be diluted with water. A Baumé hydrometer is very convenient for determining the density of the zinc solution. The latter should be maintained at from 20° to 30° in a main battery, and from 15° to 25° in a local.

It often occurs in using this battery that stalactites of copper attach themselves to the lower edge of the zinc and hang suspended in the solution, slowly but constantly increasing in length. These are first produced by a deposit of copper upon the zinc, which sets up a local action followed by a rapid decomposition of the solution and a further deposit of copper. These should be removed by means of a bent wire and allowed to fall to the bottom of the jar, as they occasion a useless expenditure of sulphate.

Absolute quietude is essential to the proper performance of this battery. A slight jar will cause the solutions to mingle, and this effect will be followed by a rapid deposition of metallic copper upon the zinc. When the zincs are removed for cleansing, care must be taken not to agitate the solution.

Prof. Hough, of the Dudley Observatory, has suggested the use of sheet lead in the place of the copper spiral, as it is cheaper and more readily cut and formed into proper shape. There is no perceptible difference in the electro-motive force or in the resistance of the battery when lead plates are substituted for copper in this way.

The electro-motive force of the gravity battery is the same as the Daniell, and the average resistance when in good working condition about three units.

164. Siemens' Universal Galvanometer.—The apparatus employed for the measurement of electrical resistances consists essentially of a standard resistance, which is used for the purpose of comparison, a galvanometer, by which the result is indicated, and a galvanic battery. In the different methods of testing, these appliances are arranged in various ways, as particular circumstances may render convenient or desirable. The various methods of testing in use may be classified, however, under three heads, viz.:

1. By the angles of deflection of a galvanometer needle.

2. By the differential galvanometer.

3. By the Wheatstone bridge, or electrical balance.

The first-named method is the simplest in principle, and, with proper care, gives very accurate results. It is not so convenient as the other two methods for ordinary use, but is applicable more especially for the measurement of very high resistances, such as insulators, etc. It is also employed in measuring the internal resistance of batteries. As the strength of the current passing through the coils of a galvanometer is always proportionate to the sine or tangent of the angle of deflection of the needle, and is also inversely proportional to the resistance in circuit, it follows that if we find the deflection with a certain known resistance in circuit to be only 22°, and we then substitute for this known resistance an unknown one, which gives us a deflection of 39°, the tangent of the latter will be twice that of the former, and the unknown resistance is consequently found to be half that of the known resistance. (170.)

The second method is very convenient and is much used, although not equal in strict accuracy to the third method. The galvanometer coils are wound with two wires of the same length and resistance, insulated from each other with the utmost care. The needle is therefore surrounded by an equal number of convolutions of each wire, which are also equidistant from it. One end of each wire is connected to the battery, but in such a manner that the current flows in opposite directions through the two wires. When, therefore, the two currents are of equal strength, one tends to deflect the needle to the right and the other to the left with equal power, and the needle remains at rest. If we insert an unknown resistance into the circuit of one of these wires the current is weakened, as is also its effect on the needle, which no longer remains balanced and at rest, but is deflected to one side. If we now insert a series of known resistances into the circuit of the other wire, until the needle is again brought into equilibrium, we are certain that the unknown resistance in one circuit is exactly equal to the known resistance in the other. (122.)

The third method is susceptible of the greatest accuracy of measurement, when proper precautions are observed. The connections of the "bridge" are arranged as follows:

We will suppose the wires A B C D (Fig. 57), arranged in the form of a parallelogram, to be of exactly equal resistance. If we attach the two poles of a battery, E, to the points 1 and 2, its current will divide at 1, half of it going through A B, and the other half going through C D, to the point 2, and thence to the other pole of the battery. The galvanometer G, placed on a wire connected across from 3 to 4, will not be affected as long as A B is equal to C D, no matter what the absolute resistance may be.

Again, when A bears the same proportion to C that B does to D, or when A: C:: B: D, no current will pass from 3 to 4 through the galvanometer. If the resistance of A be made 10, that of B 1, of C 1,000,

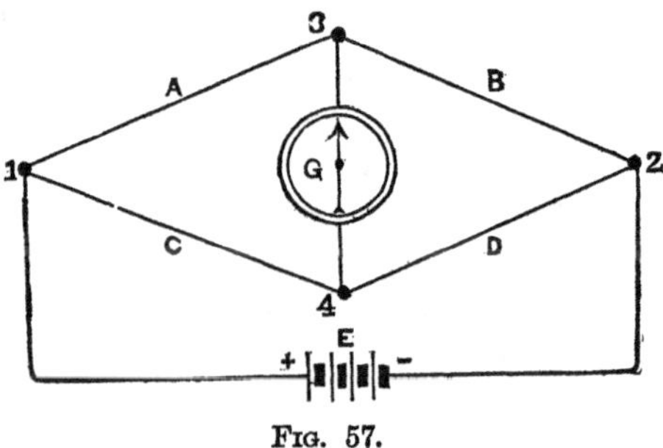

FIG. 57.

and of D 100, the total resistance of A B will now be 11, and that of C 1,100; but the tension in each branch will have fallen in the same proportion at the points 3 and 4, and no current will pass between those points.

If, therefore, we insert a known standard resistance in the wire B, and an unknown one in D, and divide a given resistance between A and C until we get no effect upon the galvanometer needle, we are then certain that the resistance of A bears the same proportion to that of B as the known resistance does to the unknown one D, which may be readily calculated by proportion or the "rule of three."

It is not necessary, of course, that the wires should be arranged in the exact form shown, nor in fact is it often done, but the principle is more easily explained and remembered when thus arranged.

165. The Universal Galvanometer of Dr. Werner Siemens is constructed upon the principle of the Wheatstone bridge, just described, but its connections are so arranged that it may be used when desired for the method of deflections first mentioned.

The galvanometer is mounted upon a disk of slate about six inches in diameter. A groove in the edge of this disk, extending about half way round the circumference, contains a wire of considerable resistance, which corresponds to the wires A and C in the above diagram. A small platina roller, mounted upon a radial arm, is connected to one pole of the battery, and forms the connection with the wire A C, as shown at 1 in the diagram. The wire corresponding to B is supplied with three standard resistances of 10, 100, and 1,000 Siemens' units, respectively, either of which may be placed in circuit at pleasure, by means of contact plugs. The wire D is provided with binding screws for the attachment of the wire, or other resistance which it is required to measure. The galvanometer consists of a pair of very delicate astatic needles, suspended by a fine silk fibre. The coil has a resistance of 100 Siemens' units.

The radial arm carrying the platina roller also carries a pointer or index, moving over a scale upon the circumference of the slate disk, which is divided into 300 degrees, and which may be read to one-fifth of a degree by means of a vernier.

In using the instrument, the standard resistance corresponding most nearly to the unknown resistance which is to be measured is unplugged and placed in circuit at B (Fig. 57), while the unknown resistance itself is inserted at D. The radial arm carrying the platina roller 1 is now moved towards A or C, until the needle is balanced. The proportion of A to C is then read off the scale, from which the proportion of B to D is readily calculated, or is taken from a printed table furnished with the instrument.

This galvanometer may also be employed for comparing electro-motive forces, according to the method of Poggendorff,* and is applicable to almost any purpose for which an apparatus of the kind may be required.

This instrument usually has a constant of about four degrees, with one Daniell's cell through 1,000,000 Siemens' units. When used as a Wheatstone bridge, its range of measurement is from 0.17 to 59,000 Siemens' units. Higher resistances, such as insulators, may be measured by the method of deflections. The entire apparatus (except the battery) occupies a space only nine inches in diameter, and the same in height. It is packed in a neat case, and can be carried about with great convenience.

166. Pope & Edison's Printing Telegraph.—Type-printing telegraph instruments, which were formerly employed for commercial telegraphing, have, within two or three years, been extensively introduced, in a modified and simple form, in the various branches of private telegraphy, with great success. One of the best of these is that of Pope & Edison, which is used on a large number of private lines in New York city.

The different portions of the apparatus, with the exception of the battery, are mounted upon a small table, similar in size and construction to that of an ordinary sewing-machine. At the back of the table are six binding screws to which are attached the line and ground wires, and the wires leading to the main and local batteries. The instrument operates upon what is known as the "open circuit principle," each station transmitting with its own main battery—the line at the receiving station being connected directly through the relay to the ground without the intervention of a second battery.

* See Clark's "Electrical Measurement," p. 105. Also Sabine's "Elect. Tel.," p. 320.

The printing apparatus is placed upon a circular iron base in the centre of the table. In front of it is placed a dial containing the letters of the alphabet, arranged in a circle and provided with an index or pointer, mounted upon a horizontal shaft. This shaft also carries a type wheel, with the letters of the alphabet engraved upon its periphery, and a scape-wheel, with rachet-shaped teeth, corresponding in number to the characters upon the type wheel and dial. An electro-magnet beneath the base is provided with an armature, attached to a vibrating lever, the latter armed with pawls or clicks, so arranged in relation to the scape-wheel that every time the electro-magnet attracts its armature, the wheel is made to revolve a distance of one tooth, and the type wheel and index upon the same shaft a distance of one letter. At the extreme right of the circular base, and partly beneath it, as seen in the engraving, is placed a second electro-magnet, whose armature lever passes in a horizontal direction below the type wheel. Directly underneath the type wheel an india-rubber pad is fixed upon the lever, by means of which an impression of the letter which is opposite it upon the type wheel may be taken in the manner hereafter to be described. This lever is also provided with a simple mechanical device for moving the paper forward the proper distance, as the impression of each successive character is imprinted upon it. This may be seen at the left of the printing apparatus. The type wheel is provided with a suitable inking roller, as shown in the engraving.

It will thus be understood that the printing mechanism is operated by two distinct electro-magnets, one of which is so arranged that its successive pulsations may be made to advance the index step by step to any required letter, while the other forces the strip of paper against the inked type upon the wheel, after it has been moved to the proper position by the first magnet. The type wheel is, of course, so arranged in reference to the index upon the same shaft, that when the latter points to any given letter the corresponding letter upon the type wheel is opposite the impression pad.

These two electro-magnets are placed in the circuit of a local battery, which is brought into action by a relay placed in the main line circuit, as in the ordinary Morse apparatus. The relay is shown at the right of the printing mechanism, covered by a small glass shade. It is the same in principle as the ordinary Morse relay, with the addition of a device termed the "polarized switch," which consists of a permanently magnetized steel bar, pivoted between the poles of the relay magnet, and forming a part of the local circuit. This is attracted to the right or left according to the polarity of the relay magnet, which itself, in turn, depends upon the direction of the electrical current in the main circuit. The polarized switch determines the direction of the local circuit, causing it to pass through the magnet for moving the type wheel, or through the impression magnet, as may be required.

Two lever finger keys, with vulcanite knobs, are placed on each side of the printing apparatus, as shown in the engraving; and it is by means of these that the instrument is operated. They are connected to opposite poles of the main battery in such a manner that, by depressing the right hand key, the positive pole of the battery is connected, through the relay magnet to the line, and the negative to the ground, while the left hand key, on the contrary, sends a negative current through the relay and line in the same manner.

The mode of operating the instrument is exceedingly simple. By depressing the right hand key a sufficient number of times in rapid succession, a series of positive currents is sent through the relays at both ends of the line, which are repeated upon the local circuits of both instruments. The positive currents deflect the polarized switches to the left, so that the local circuit is directed into the type-wheel magnet. The index and type wheel of both instruments, therefore, advance one letter every time the key is depressed, and they may thus be readily brought to any desired letter. When this has been done, the left hand key is depressed, which sends a negative current, reversing the polarized switch, and the local circuit is directed through the printing magnet, producing the impression of that letter upon the strip of paper, and this process may be continued indefinitely.

Suitable arrangements are provided for bringing the type wheels of the two instruments together in case they should accidentally be thrown out of correspondence.

These instruments are entirely automatic in their action, and a despatch may be printed at the remote end of the line, in the absence of an attendant. In the event of any derangement of the printing apparatus, it may be used as a dial instrument as conveniently as if especially constructed for that purpose.

The battery is always disconnected, except at the moment of working, and therefore is consumed but slowly. Other systems require the battery to be constantly connected to the line whether working or idle. A battery of two carbon cells per mile, and in many cases even less, will work the instrument and remain in action from one to four weeks without renewal, according to the amount of telegraphing done upon the line.

It is impossible for the main circuit to be accidentally left open. Only one adjustment—that of the tension spring of the relay—is required after the instrument is first put in operation, and that but rarely on lines of ordinary length.

CHAPTER X.

APPENDIX AND NOTES.

167. THE EQUIPMENT OF TELEGRAPH LINES.—The satisfactory performance of any given telegraphic circuit depends largely upon the maintenance of a proper relation between the respective resistances of the line, instruments, and batteries. There is in all cases an ascertainable definite proportion between these, which gives, theoretically, the best result with the least expenditure ; to which practice should always be made to approximate as nearly as possible. The disregard of the well-established laws of electrical and magnetic action is not only the source of grave difficulties in the practical operation of lines, but also entails an enormous waste of material and supplies.

It is one of the fundamental laws of the electric circuit, that *with a given resistance of conducting wire and battery, the maximum magnetic force is developed when the total resistance of the coils of the electro-magnet or magnets is equal to the resistance of the other portions of the circuit, i. e., the batteries and conducting wires.* (173.)

The resistance of the conductor, which must of necessity, always form a large proportion of the total resistance in every main circuit, is in practice determined within certain well-defined limits, by considerations of distance, mechanical construction, and first cost. It therefore becomes necessary to adjust the resistance of the remaining parts of the circuit with reference to that of the conductor, which in practice usually ranges from 10 to 20 units per mile. With the

No. 9 galvanized iron wire generally used, it approximates closely to the latter figure.

The resistance of the batteries forms but a very small portion of the total resistance in an ordinary main circuit, and admits of comparatively little variation, so that the actual problem which presents itself, is to determine the proper resistance of the relays when the resistance of the conductor is given, and the form of battery which will supply the necessary electrical power for operating the line with the least expenditure of materials and labor.

The size of the conductor having been fixed upon, this taken in connection with the length will determine its total resistance. The combined resistance of the relays should be made to equal this amount as nearly as possible. It is hardly necessary to add that the resistance of the different relays should be uniform in respect to each other. With good relays the amount of battery required to operate the main circuit should not exceed 1 cell of Grove or Carbon battery for each 150 units resistance, and will generally be less than this. About double this number of the Daniell, Hill, or Callaud battery will be needed.

For example, suppose it is required to construct a telegraph line 300 miles in length, with 15 stations. If No. 9 iron wire is used as a conductor its resistance will be say $300 \times 20 = 6000$ units. The resistance of all the relays being made equal to that of the line, we have as the proper resistance for each relay $\frac{6000}{15} = 400$ units. The amount of battery required will be $\frac{12000}{150} = 80$ cups of Grove or Carbon, or about 160 cups of Daniell, Hill, or Callaud.

The approximate average resistance, and comparative electro-motive force of the different batteries in use is as follows, the Grove battery being taken as the standard at 100 :

	Resistance.		Electromotive force.
Grove	.5	units.	100
Bi-Chromate or Carbon	1.0	"	107
Daniell	2.0	"	56
Callaud	3.0	"	56

These figures refer to the ordinary sizes of the Grove and Carbon battery, and to the Daniell and Callaud when adapted to a jar eight inches high and six inches inside diameter. Although the resistance of the battery when included in a single main circuit of the usual length, has but little influence upon the effective strength of the current as a whole, yet in local circuits, and in main batteries from which a number of lines are worked at the same time (110), it becomes an essentially important element in the calculation.

Another important law of electrical action, which applies especially to instruments which are to be worked by a local circuit, is the following :

The greatest effective force of any given battery is developed when the sum of all the external resistances in the circuit is equal to the internal resistance of the battery.

In a local circuit there are practically no resistances except those of the battery and magnet, and it is therefore obvious that these should be so adjusted as to equal each other as nearly as possible. Tested by this rule, a great portion of the sounders, registers, and repeaters, in use in this country, will be found to have magnets of too low resistance, most of them being adapted to the use of a local of 1 Grove cell, although nearly all the local batteries in use are composed of 2 or 3 cells of Daniell. Such a magnet will only partially develop the effective force of a Daniell battery, and still less that of a Callaud or Hill.

The sizes of copper wire generally used in local helices vary from No. 19 to 22, American gauge, and the resistance from 0.5 units to 4 units. The most usual resistance is about 1 unit. If we take a sounder of this resistance and apply a cell of Grove battery, we have the following result :

Resistance of magnet....................	1 unit.
" " battery....................	1 "
Total.............	2 "

Calling the electro-motive force 100, and dividing this by the resistance, we get 50 as the effective strength. If we take the same sounder and apply a Daniell element with 2 units resistance the total resistance will be 3, the electro-motive force 56, and the quotient or effective force 18.6, but little more than one-third that of the Grove. With 2 Daniell cells we have—

Resistance of magnet	1 unit.
" " battery	4 "
Total	5 "

The electro-motive force of 2 cells will be $56 \times 2 = 112$, and dividing this by the resistance, 5, we have 22.4. With 2 Callaud cells the effect would be still less, in fact only 16.

Now let us take the same sounder, and remove the helices of No. 19 wire, which give a resistance of 1 unit, and rewind them with No. 23 wire, and observe the effect. With a given strength of current, the magnetic effect is proportional to the number of convolutions, and the latter increase inversely as the square of the diameter of the wire. The resistance of the wire also increases as its length, and inversely as the square of its diameter. The squares of the respective diameters would be as follows:

No. 19	.00128881
" 23	.00051076

The average length of each convolution in a helix of a given size will be the same with any sized wire. The length being in inverse proportion to the square of the diameter, the resistance due to the increased length will be

$$.00051076 : .00128881 :: 1 \text{ unit} : 2.52 \text{ units}.$$

But the resistance is further increased in inverse proportion to the *square of the diameter* of the wire, therefore

$$.00051076 : .00128881 :: 2.52 : 6.3$$

6.3 units would, therefore, be the resistance of the new helices. This is not strictly accurate, as no allowance has been made for the spaces between the convolutions, which occupy more room in the coil when finer wire is used, and somewhat reduce the number of convolutions as well as the length and resistance of the wire. We will, therefore, call the resistance of the new helices 6 units. This resistance will give the greatest possible effect obtainable with 2 Callaud cells, which will be as follows:

Resistance of magnet........	6 units.	
" battery.	6 "	
Total resistance...	12 "	
Divide the electro-motive force	112	= 9.3
By the total resistance.......	12	

But the magnetic effect is increased by the greater number of convolutions in the proportion of the squares of the diameters, or as 2.52 to 1. Therefore $9.3 \times 2.52 = 23.43$. This is greater than the magnetic effect of 2 Daniell cells upon the sounder of 1 unit resistance, which we before found to be 22.4. Making some deduction for the slight decrease in the number of convolutions, owing to the greater number of spaces, we may consider the actual magnetic effect to be the same in both cases. Experience has shown that this is amply sufficient to operate a well-constructed sounder or register.

In the above calculations the resistance of the Daniell is given as 2 units. It is actually over 3, except when the porous cell is defective, or so excessively porous as not to separate the liquids properly. The Callaud is also given as 3 units, but in point of fact does not exceed 2 after it has been 2 weeks in use. The resistance of the different Callaud cells is very uniform, while cells of the ordinary form of Daniell will often vary widely under precisely similar conditions. Sometimes one cell will measure 10 units, and another only 2, owing principally to difference in the quality of the porous cups. A cell of high resistance will diminish instead of increasing the effect in a local circuit.

The obvious advantage of using the Callaud battery for local circuits in connection with a magnet whose resistance is properly adjusted to it, consists in its great economy, the expense of maintenance not being more than one-fifth as great as when the ordinary Daniell is employed. The above calculations show that a great saving can be made when the Daniell itself is used, by regulating the resistance of the magnets to correspond with that of the battery.

168. THE WORKING CAPACITY OF TELEGRAPH LINES.—In order to secure the best possible result in the working of telegraph lines we must keep down the resistance of the conductors in the circuit (42), and increase the resistance of the insulation (90) to the greatest practicable extent. In other words, the resistance must be as small as possible in the route we wish the electric current to travel, and as great as possible in every other direction. *The practical working value of a telegraph line is the margin between the joint resistance of the conductor and the insulation, and that of the insulation alone.* The tension of the retracting spring of the relay armature, when upon a "working adjustment," is the measure of this margin or difference. It is evident that this margin may be increased in two ways, viz.:

1. By increasing the insulation resistance.
2. By decreasing the resistance of the conductor.

For example, suppose a line of telegraph 100 miles in length—the weather being rainy. Suppose that the conductor has a resistance of 20 units per mile, while the resistance of the insulators is 1,000,000 units per mile. Let the receiving magnet and battery be situated at one extremity of the line and the key at the other. When the key is closed, the force acting upon the armature of the magnet is in proportion to the quantity of electricity leaving the battery and passing through the magnet to the line, and this quantity is made up of that escaping through the insulation along the line, in addition to that going through the conductor to the other end of the route. When the key is open, the force exerted upon the armature is due to the current passing through the insulation alone. The effective working strength is therefore the difference between the attractive forces acting upon the armature, when the key is opened, and when it is closed at the other end of the line—or, in other words, *the working margin is the difference between the sum of the forces due to the joint conductivity of the wire and insulators and that of the insulators alone* (104).

Thus, in the case cited:

The total resistance of the wire is....................	2,000 units.
" " insulation....................	10,000 "
The joint resistance of wire and insulators is	1,666 "

The strength of current being inversely proportional to the resistance, it will be as follows:

When key at other end is closed......................	100.00
" " " open......................	16.66
Difference, or effective working margin.............	83.33

It is not the *absolute* resistance of the conductor or of the insulators that determines the value of a line. It is operated by the *margin or difference* between these two values (101). It is important that this should not be lost sight of.

Now let us observe the effect of substituting a wire of twice the weight, having a resistance of only 10 units per mile. We now have:

Total resistance of wire............................	1,000 units.
" " insulation (as before)...........	10,000 "
Joint resistance.......	909 "

The proportionate strength of current will become :

When key is closed..............................	100.00
" " open..................	9.09
Difference....	90.91

We have given the strength of current with key closed as 100 in both the above cases, in order to show the proportionate increase of margin. The *absolute* strength of current in the two cases is as 100 to 183, an increase of 83 per cent., while the increase of working margin is only 9 per cent.

We will now take the result of an actual measurement. A new No. 9 galvanized wire, 115 miles in length, on a clear and fine day, gave a resistance of 2,400 units, or about 21 units per mile. On the same poles was a No. 10 plain wire, which had been in use nineteen years. This wire, including eight instruments in circuit, gave a resistance of 13,300 units. In a rain the insulation resistance of the good wire measured 15,300 units, and the bad wire 19,650.

The joint resistance of the good wire and its insulators was 2,077. The proportion of current escaping by the insulators was to the whole current as 13.51 to 100, giving a margin to work on of 86.49.

The joint resistance of the bad wire and its insulators was 7,982. The proportion of escape to the whole current was as 40 to 100, giving but 60 per cent. as an available working margin. This wire could not be worked except when the other circuits on the same poles remained idle, either closed or open. The good wire was worked without difficulty. The escape was apparent, but was not sufficiently great to cause any serious inconvenience. The relative working margins were in the proportion of 86.49 to 60.

On a clear and cold day the insulation of the good wire showed a resistance of 2,400,000 units, the working margin being 99.99. The bad wire showed an insulation resistance of 1,700,000 units, the working margin being 99.93. The difference in this case between the two wires was only 00.06, an amount not appreciable in practice. The poor wire worked as well as the good one, but the current was not so strong. This difference could be compensated for by increasing the battery on the former.

In the above instance we have two wires on the same poles. One is new and a good conductor, the other old and a poor conductor. In fine weather the insulation of the new wire is the most perfect, but the difference in their working is inappreciable. In rain, although the insulation of the old wire is actually the best, yet it does not work nearly so well as the new wire, and this is attributable solely to the fact that the new wire has a much greater conductive capacity.

Take another example, also from actual measurement: A new wire, 150 miles in length, on a clear day gave a resistance of 2,200 units. On the same poles was an old rusty No. 11 wire, which gave a resistance of 23,500 units. On a very wet day the insulation resistance of the new wire was 4,800 units, and of the old wire 32,000 units. The working margin of the new wire was 78, and that of the old wire 60. In this case the amount of current escaping over the insulators of the new wire was 2.7 times that passing through the old wire and its insulators combined! In other words, the current with key open on the new wire was nearly three times as strong as on the old wire when the key was closed.

In these examples the resistance of the batteries and instruments has not been taken into account, as they do not materially affect the results.

169. The Electrical Tension of Telegraphic Batteries and Lines.—In another part of this work (8) it was briefly stated that the electrical tension of a battery, or its power of overcoming resistance, is increased in direct proportion to the number of elements of which the battery consists. Suppose we have a battery of 100 cells, and the electro-motive force of each element

of this battery be such as to produce a difference in tension between its plates equal to 1, the difference between its poles or end plates will be equal to 100. But it must be understood that degrees of tension are only relative or comparative. The earth being our great reservoir of electricity, its tension is called zero, and it affords us a convenient standard of reference in comparing other tensions, but even the absolute tension of the earth sometimes varies in different times and places.

Suppose we take the battery of 100 cells above referred to, place it upon a well-insulated stand, and connect one pole of it, say the zinc or negative pole, to earth, and leave the other pole disconnected, and therefore insulated by the air. The end which is connected with the earth being in free communication with it, will now have a tension of zero, and the opposite end of the battery will have a tension of 100 positive, or above that of the earth, and if a wire were connected from it to the earth a powerful current of electricity will pass between them.

If now the copper or positive pole be placed to the earth, and the zinc pole insulated, the tension of the former will now be zero, and that of the latter 100 negative, or *below* that of the earth. In each of these cases the degree of tension is the same, but in one case it is above that of the earth, or positive, and in the other case below that of the earth, or negative.

If the zinc or negative pole of the same battery be now connected to the earth, and the positive pole, instead of being left free, is connected by a short and thick wire, of no appreciable resistance, to the negative pole, the tensions throughout the circuit will be materially changed, although the electro-motive force will remain unaltered. The tension at the copper pole of the battery, which was 1,000 when the pole was entirely disconnected, now becomes the same as that of the earth, or at least but very little above it. If a wire offering considerable resistance be substituted for the short and thick wire which connects C and Z, the tension at C will be raised, although that of Z will still be kept at zero by its connection with the earth at that point. In proportion as the resistance of this connecting wire is increased, the tension at C rises until, when the resistance becomes infinite, the tension will again reach 100, for infinite resistance is absolute insulation. The tension is now equal to the electro-motive force, but it is obvious that it can never exceed it under any circumstances.

If a battery of 100 cells is connected to a telegraph line of 100 miles in length, whose insulation is perfect, and which is not connected to the earth at the remote end, the line will instantly acquire a tension of 100 throughout its whole length (this being equal to the electro-motive force of the battery), and this would occur if the wire were a thousand or a million times that length. After the line has acquired the same tension as the pole of the battery to which it is attached, no current will flow from the battery.

If the distant end of the line is connected to the earth, the battery will come into action, and a current of electricity will pass through it. This will at once change the tensions throughout the whole line. The distant end of the line, which originally had a tension of 100, will now have a tension of zero, being connected directly to the earth, and from this point the tension will rise gradually and regularly along the whole length of the line to the pole of the battery. So also the tensions within the cells of the battery itself follow the same law.

The relation existing in a voltaic circuit between the resistances, electro-motive forces, and tensions, may be graphically and accurately represented to the eye by a geometrical projection based upon mathematical reasoning, a method first suggested by Ohm, and more recently elaborated by Mr. F. C. Webb, and which he explains as follows:

Let all the parts of a circuit, whether liquid or solid,

be expressed in their successive order by portions of a continuous horizontal line, which shall be to one another as the reduced lengths or resistances of those parts. Let the tension at any given point in the circuit be represented by the perpendicular height of a point above, or depth below, the horizontal line representing the resistances. This when above the line will indicate a positive, and when below, a negative tension. The horizontal line of resistances may be termed the axis.

In order to represent the tension at every point in the circuit, we must construct a line termed the *line of tension.* The perpendicular height of this line above the axis at any point, indicates a corresponding positive tension at that point, and its depth below in the same manner indicates a negative tension. When this line crosses the axis the point of intersection has no tension.

Electro-motive force consists in a sudden and constant difference in the tension of the points situated immediately upon opposite sides of the surface of junction between the zinc element and the liquid of the battery. The electro-motive forces in the circuit must, therefore, be represented by a sudden rise in the line of tension at the points along the axis at which they occur, thus forming lines perpendicular to the axis. The magnitude of these lines must be proportional to the electro-motive force they represent. Moreover, as the electro-motive force is a quantity depending solely upon the nature of the elements at the surface of junction at which it occurs, and not at all on any change in the resistance or electrical state of the circuit, these perpendicular lines constantly maintain the same magnitude, although their position as regards the axis may be altered in various ways.

Now let us construct a diagram which shall correctly represent the electro-motive forces, tensions, resistances, and strength of current, as a telegraph line with a closed circuit, having a battery of three cells at each end of the line, which will be a sufficient number to correctly represent the arrangement of the circuit ordinarily used on American telegraph lines.

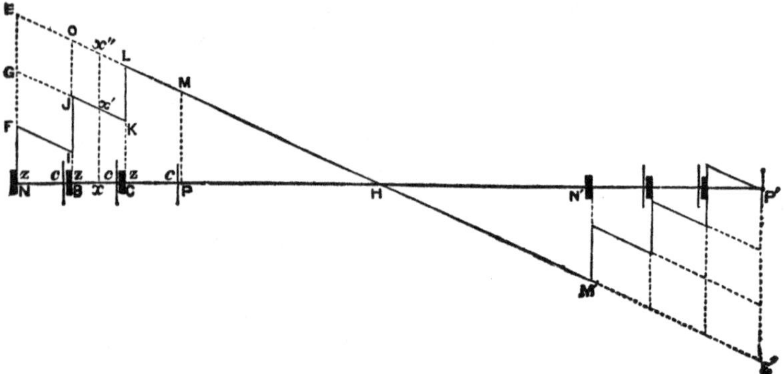

Let the horizontal line N P′ (see Fig. 59) represent the axis, or line of resistances, the latter being represented in their respective order, beginning at the point of contact, N, between the extreme zinc plate of the battery and the liquid of the cell. N B, B C, and C P represent the respective internal resistances of the three battery cells, and N P that of the entire battery. Let P H N′ represent the resistance of the line wire, and N′ P′ that of the battery at the opposite end of the line. Erect a perpendicular, N E, at the point N, and divide it into three portions, N F, F G, and G E, which shall be to each other as the electro-motive forces at N, B, and C. The other battery, N′ P′, having its negative pole, N′, to the line, will give a *negative tension;* therefore a perpendicular P′ E′ let fall below the axis from the point P, and divided in the same manner, will represent the electro-motive forces of the battery N′ P′. Therefore the line N P′ represents the sum of all the resistances, and N E + P′ E′ the sum of the electro-motive forces. It necessarily follows that the line of tension, M H M′, which we get by joining E and E′, varies in the angle of its inclination to the axis according to the proportion between the sum of the electro-motive forces, N E and P′ E′, and the sum of the resistances, N P′; and the *degree* of its inclination will therefore accurately represent the effective working strength of the current in all parts of the circuit.

The varying tensions within the battery may be correctly represented as follows: Having joined E and E′, erect perpendiculars at B and C and P. Now as the effective strength of current, represented by the *inclination* of the line E E′, is the same at every point throughout the whole circuit, draw F I parallel to E E′. Then F I will be the line of tension in the first cell, falling regularly through the resistance of the liquid to the surface of generation, B, of the second zinc, where it rises suddenly to the extent of the electro-motive force there situated. Draw G K parallel to E E′, intersecting B O at J. I J will then be equal to F G, which represents the electro-motive force at B, and J K will be the line of tension in the second cell. Now as G K is parallel to E E′, K L will be equal to G E, the electro-motive force at C, and L M will be the line of tension in the third cell. In the same manner the line of tension within the other battery N P′.

The terminal points of the line N and P′, being connected directly with the earth, their tension will be equal, and the same as that of the earth, which is assumed to be zero; that is, neither positive nor negative. It is manifest that at the point H, midway of the circuit where the line of tension crosses the axis, the tension is the same as that of the earth, or zero.

In the illustration given, the line is supposed to be in a condition of perfect insulation. In actual practice there is a leakage at every support throughout the whole length of the circuit. The line of tension in this case would form a double catenary curve, its angle of inclination to the axis constantly increasing from H to M and M′, because in an imperfectly insulated or leaky line the current continually increases in strength in each direction from the neutral point to the battery poles at P and N′.

Mr. Webb has demonstrated the correctness of the above method of geometrical projection by applying Ohm's formula for obtaining the tension at any point of the circuit. The results are found to correspond in every case. This formula may be stated as follows:

Let T = the tension at any given point of the circuit x.
Y = the abscissa of that point x, taking as origin the point of least tension.
A = the sum of the electro-motive forces.
L = the reduced length or resistance of the entire circuit.
O = the sum of the electro-motive forces included in Y.
C = the tension of the whole circuit to external objects. That is to say, the tension of the circuit, if it be an insulated circuit, and electrified by a source not contained within it.

$$\text{Then } T = \frac{A}{L} Y - O + C.$$

As, in the case under consideration, the earth forms part of the circuit, the constant C disappears and the formula becomes

$$T = \frac{A}{L} Y - O.$$

Now take a point x in the diagram, and the tension $x\,x'$ will be found to agree with the formula.

The quantities in the formula are thus represented geometrically in the figure:

$$A = N\,E$$
$$L = N\,H$$
$$O = x'\,x''$$
$$Y = x\,H$$
$$T = x\,x'$$

Now since the triangles H N E and H x x'' are similar, we have

$$N\,H : N\,E :: H\,x : x\,x''$$

$$\text{Consquently } x\,x'' = \frac{N\,E}{N\,H} H\,x,$$

and J K being parallel to O L, we have

$$x'\,x'' = K\,L.$$

$$\text{But} \qquad x\,x' = x\,x'' - x'\,x'';$$

$$\text{Therefore } x\,x' = \frac{N\,E}{N\,H} H\,X - x'\,x'',$$

$$\text{Or,} \qquad T = \frac{A}{L} Y - O.$$

An experimental proof of the above theory of tension may be obtained by connecting a wire from the neutral point in the middle of the closed circuit of a telegraph line, and inserting a galvanometer or relay. It will be found that no current passes between the line and the earth, which proves that the electric tension or potential at that point is zero, or the same as that of the earth itself.

170. Double Transmission.—One of the most interesting problems in practical telegraphy is that of double transmission, or working in opposite directions at the same time over a single wire. This apparently paradoxical result may be accomplished in several different ways, the principles involved being very simple and easily understood. The method shown in the accompanying diagram is that of Siemens & Halske, of Berlin, Prussia; the apparatus now used in this country differing slightly from it in some of its minor details.

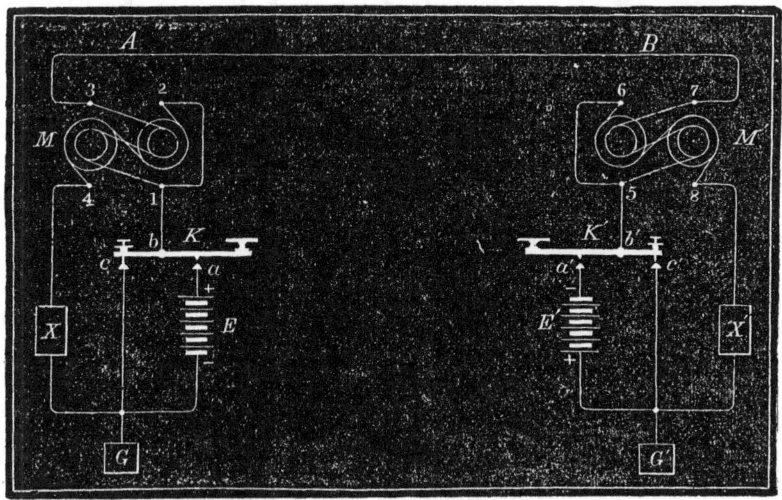

Fig. 60.

A and B (Fig. 60), are the two terminal stations of the line. The main battery E, at station A, is placed with its +, and the battery E′ at station B with its —

pole to the line, as represented. M and M′ are the receiving magnets or relays, which are wound throughout with two similar wires of equal length, as shown in the figure, whose connections will hereafter be explained. The rheostat or resistance, X, must be adjusted so as to be exactly equal to that of the line A, B, added to that of the relay wire 7, 5, at the other station. Similarly X′ is also made equal to the line including the relay wire 3, 1.

If, now, the key K at station A be depressed, the current from the battery E will divide at the point 1, one portion going through the relay coils to 3, over the line A, B to 7, and thence through the relay M′ to 5, key lever 6′, and contact C′ to the earth at G′, and the other portion in an opposite direction through the relay coils from 2 to 4, and thence through the resistance X to the negative pole of the battery. These two currents will be equal to each other, the resistance being the same by each of the two routes, as before explained, but as they pass in opposite directions through the two wires surrounding the relay M, they produce no magnetic effect upon it. The relay at B, however, will be affected by the current coming from A through the wire 7, 5, and will give signals corresponding to the movements of the key at that station.

If, now, the key at B be also depressed, the same action takes place; one half the current passes over the line, combining with the current from A, and the other half returns to the battery through the other wire of the relay and the rheostat.

The relay wires 1, 3 and 7, 5 are now traversed by the double current, equal to $\frac{A}{2}+\frac{B}{2}$, but the wires 2, 4 and 6, 8 are traversed only by the current of a single battery, having at A the force of $\frac{A}{2}$ and at B the force of $\frac{B}{2}$. The latter current being in the opposite direction to the former, the relays at both stations are affected by the difference in the forces of these currents, the relay at A by $(\frac{A}{2}+\frac{B}{2}) - \frac{A}{2}$, and the relay at B by $(\frac{A}{2}+\frac{B}{2}) - \frac{B}{2}$.

Thus each station receives its signal through the action of the distant battery only.

In the arrangement shown in figure 60 a third position occurs, where one of the keys, at B for instance, is in the act of changing from the front contact A′ to the rear contact C′, or *vice versa*, in which case the current from A is interrupted at B′, and therefore passes through the second wire of the relay 6, 8, but this time in the same direction, and thence through the rheostat X′ to the ground. The current arriving at B is considerably weakened in consequence of the additional resistance encountered at X′, but this is compensated for by its passing through *both* wires of the relay M in the same direction, and its action upon the relay, therefore, remains about the same as before.

One slight difficulty, however, arises in this connection. It will be seen that when the current at the receiving station is thus momentarily thrown through both relay wires and the rheostat, it must necessarily cause an unequal division of the current between the two opposing relay wires at the sending station, as the resistance of the long circuit becomes about double that of the short one. This effect is avoided in the American system by a modification of the transmitting apparatus, which is operated by the lever of a sounder placed in a local circuit in connection with the key. When the local circuit is closed the downward movement of the sounder lever makes the battery connection upon a flat spring, and the movements thus imparted to the spring breaks the earth contact. The spring being attached to the line wire the connection is necessarily always complete, either direct or through the battery, and it is not obliged to pass through the rheostat when the transmitter is changing from the battery to the earth contact, or *vice versa*. The disadvantage in this case arises from the fact that the main battery is thrown on short circuit at each movement of the transmitter, rendering it necessary to interpose a considerable additional resistance between the back contact and the battery, to prevent the rapid consumption of the latter which would otherwise ensue. These improvements were devised by Mr. J. B. Stearns.

In working this system, it is necessary to keep the rheostat so adjusted that its resistance will correspond exactly with that of the line, as above shown. If the relay works too feebly the counter current must be weakened by increasing the resistance of the rheostat. If the magnetism is too strong the resistance should be diminished. A careful study of the diagram will show that this system operates equally well, whether similar or opposite poles of the two batteries are placed towards the line. With like poles the action will be as follows:

If the key at A be depressed, the current on the line will be $\frac{A}{2}$ and through the rheostat $\frac{A}{2}$, neutralizing each other upon the relay of A, but giving a current of $\frac{A}{2}$ in the relay at B. Now, if the key at B be also depressed, a current equal to $\frac{B}{2}$ is thrown through each wire of his relay, but the current $\frac{A}{2}$ being equal and opposite to $\frac{B}{2}$ the current of the main line will $= 0$.

The current through the second wire of the relays being still unaffected, each relay will give a signal corresponding to the time the key at the other station is depressed.

171. Edison's Button Repeater.—This is a very simple and ingenious arrangement of connections for a button repeater, which has been found to work well in practice. It will often be found very convenient in cases where it is required to fit up a repeater in an emergency, with the ordinary instruments used in every office. Fig. 61 is a plan of the apparatus.

M is the western and M′ the eastern relay. E is the main battery, which, with its ground connection G, is common to both lines. E′ is the local battery, and L the sounder. S is a common "ground switch," turning on two points, 2 and 3. In the diagram the switch is turned to 2, and the eastern relay, therefore, repeats into the western circuit, while the western relay operates the sounder, the circuit between 1 and 2 through the sounder and local battery being common to both the main and local currents. If the western operator breaks the relay M opens, and consequently the sounder, L, ceases to work. The operator in charge then turns the switch to 3, and the reverse operation takes place; the western relay repeats into the eastern circuit, and the eastern relay operates the sounder. The sounder being of coarse wire, offers but a slight resistance to the passage of the main current.

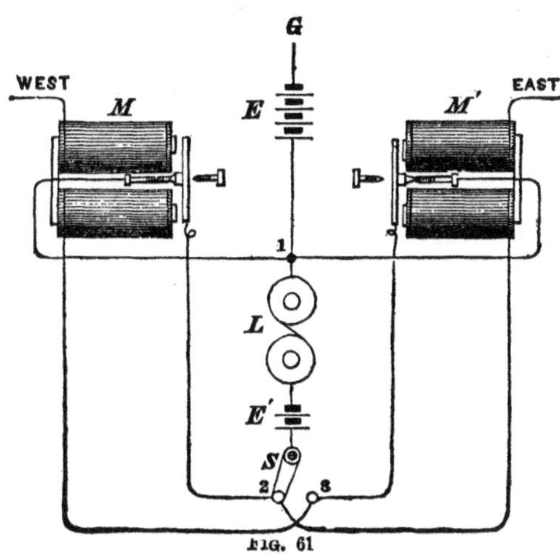

FIG. 61

172. BRADLEY'S TANGENT GALVANOMETER.—The common galvanometer used for the measurement of electric currents consists of a magnetized steel needle, suspended in the centre of a hollow frame covered with insulated copper wire. The degree of deflection of this needle from its normal position in the magnetic meridian, when a current is passing, indicates the strength of the current. In the ordinary galvanometer, however, the angle through which the needle is moved, or in other words, the number of degrees over which it passes, is not an accurate measure of the strength of the current when the deflection exceeds 15°, for the further the needle moves from a position parallel to the wires of the coil the more nearly does it approach a right angle, in which position the effect is null, so that the action of the current upon it becomes less and less powerful as the deviation increases. Several arrangements have been tried in order to obviate this objection, the most common being that of a ring having a groove on its edge

filled with wire. The needle is hung precisely in the centre of the ring, and must not be longer than one sixth of its diameter—a half inch needle requiring a three inch ring. The needle is then deflected with a force varying as the *tangent* of the number of degrees through which the needle moves. Owing to the great distance of the coil from the needle, this arrangement has very little sensitiveness compared with the common galvanometer.

In Bradley's Galvanometer a compound needle is employed, composed of several needles of thin, flat steel, fixed horizontally upon a light flat ring of metal, forming a complete circular disc of needles, having an agate cup in the centre, to rest upon the pivot upon which it moves. At each extremity of the meridian light points project, to indicate the degrees of deflection. This compound needle, after having been magnetized, is placed within or over a coil whose breadth is exactly equal to the diameter of the disc. This compound circular needle, being under the influence of the same number of convolutions of the coil in all its deflections, fulfils the required conditions for a true tangent galvanometer.

The theorem, "*The intensity of currents, as measured by the tangent galvanometer, is proportional to the tangents of the angles of deflection,*" may be verified in the following manner:

Call the terrestrial magnetism, whose tendency is to direct the galvanometer needle to the magnetic meridian, the unit of directive force, and let this unit be represented geometrically by the line A M (Fig. 62), which is the radius of the circle M B M—the line M A M representing the meridian. When there is no other force acting on the needle its direction is with the meridian. Now let an electric current be sent through the galvanometer coil, whose directive force is precisely equal to the terrestrial force, and whose tendency is to direct the needle in a line perpendicular to the meridian, and let this force be represented by the line A B.

If the terrestrial force could now, for a moment, be suspended, the needle would point due east and west; but the combined action of the two equal forces will direct the needle toward the point of intersection of the line drawn perpendicularly from M, and that drawn horizontally from B, at 1, which direction cuts the quadrant at 45°, the line M 1 being the tangent of 45°, which is 1.

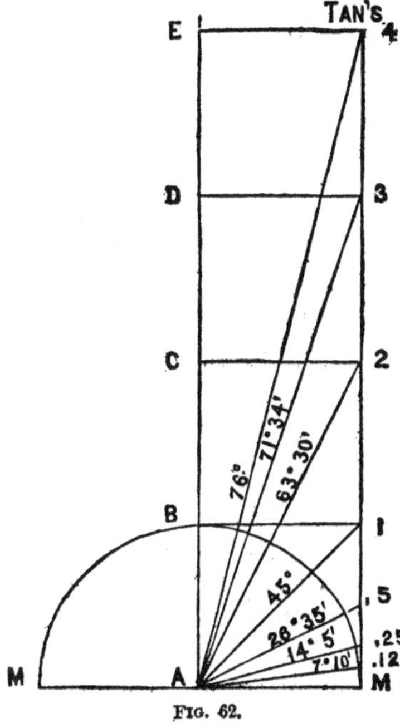

FIG. 62.

Now, if we augment the intensity of the current through the coil to twice its present force, which will be 2, and will be represented by the line A C, the combined forces A M and A C will direct the needle toward the point 2. If we now lay a protractor on the circle, we find that the line A 2 cuts it about 63° 30′, of which the tangent is 2.

We may increase the parallelogram erected upon A M at pleasure, and the two forces combined will always so balance the needle between them as to make it point from A, diagonally, across the parallelogram to its opposite angle, the height of which is the tangent of the angle of deflection.

By inspection of the diagram it is seen that the law holds good in the subdivisions of the force A B, as at .5 .25 and .125, a truth admitted by all experimenters, as to the relations, up to 14°.

173. THOMPSON'S REFLECTING GALVANOMETER.—This is the most delicate apparatus of this kind which has yet been devised, and is for this reason employed in operating the Atlantic Cables.

The special feature which distinguishes this galvanometer from an ordinary one, is the extreme lightness of the magnet or needle, and the delicacy with which it is suspended in a horizontal position. Instead of an index needle, to render the motions of the magnet visible to the eye, a reflected ray of light is made use of, which, of course, can be made of any required length. This arrangement is of great practical value in measuring faint electrical currents, too feeble to be indicated by any other apparatus. It is especially valuable in submarine telegraphing, because it permits the use of such extremely low battery power.

When the insulation of a cable is in the slightest degree defective at any point, a current of intensity has a tendency to aggravate the fault, and to corrode and eat away the conductor by chemical decomposition, at the point where the escape occurs, finally destroying the communication altogether.

Fig. 63 is a side elevation of this instrument, showing a section through the galvanometer coils and the outer case containing them. Fig. 64 is a cross section through the coils, showing the magnet, technically termed the needle. Similar letters refer to like parts in both figures. The magnet A is a small bar of steel, one half inch in length and one tenth of an inch square, cemented to the back of a very thin circular glass mirror, *a*. The mirror is suspended in a brass frame, B (Fig. 64), by an exceedingly delicate silk fibre, and is adjusted in height by the screw *b*. This frame slides into a vertical groove left in the centre of the coil, dividing it into two parts. The coil and mirror are enclosed in the brass case D, this case having pieces of glass let in wherever necessary, to permit the passage of light. The object of this arrangement is to prevent the mirror and its attached needle from being disturbed by currents of air.

A narrow pencil of luminous rays from the lamp, E, passes through the opening, F, which is capable of adjustment by the slide G. This pencil of light, passing through the lens, is reflected by the mirror back through the lens upon an ivory scale at I, as shown by the dotted lines. The scale is horizontal, extending to the

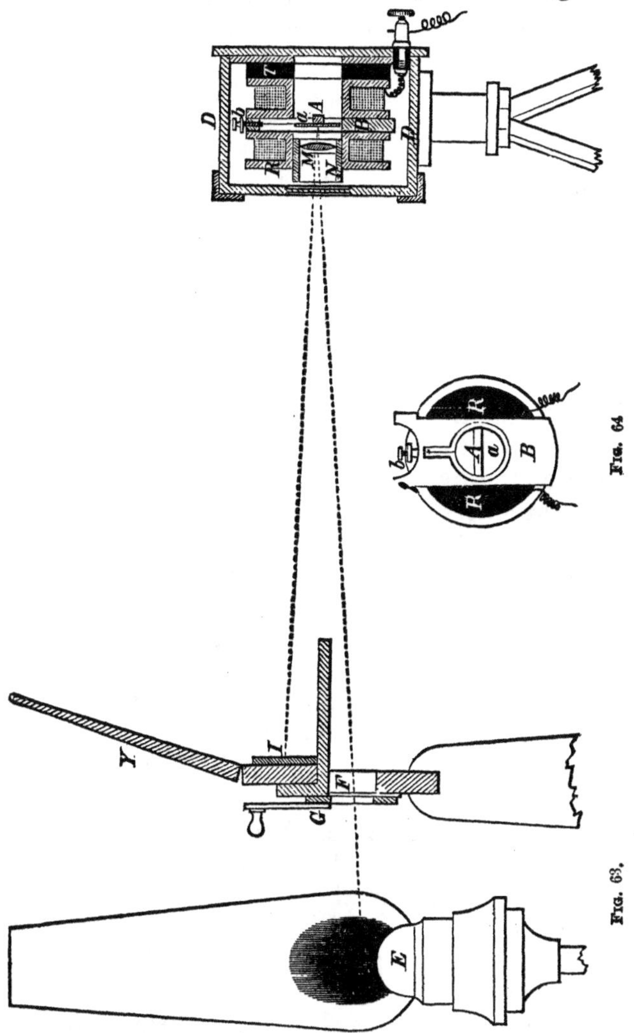

Fig. 63.

Fig. 64

right and left of the centre of the instrument, the zero point being exactly opposite the lens. The luminous pencil is brought to a sharp focus upon the scale by a sliding adjustment of the lens M, in the tube N. When the needle is at rest in its normal position, and no current is passing, the spot of light which serves as an index will remain at zero on the scale.

The operator reads the signals from a point just in the rear of the magnet and coils, the light of the lamp being cut off by the screen Y, so that he only sees the small luminous slit through which the light enters the instrument, and a brilliantly defined image of the slit upon the white ivory scale just above, which is kept in deep shadow by the screen Y. A very minute displacement of the magnet gives a very large movement of the ray of light on the scale I, the angular displacement of the ray of light being double that of the needle.

It is obvious that the ray of light from the needle will be reflected to the right or left of zero on the scale, according as the deflection is produced by a positive or negative current. The Morse alphabet is used for signaling through the Atlantic cable, deflections on one side of zero indicating dots, and on the other side dashes.

It will be observed that the end, and not the broad part of the flame of the lamp, is presented to the slit F, which is also arranged to receive the brighest part of the vertical section of the flame.

The galvanometer coils, R, consist of many thousand convolutions of fine insulated copper wire, and they are insulated from the case, D, by a disc of hard rubber, T, to which they are fastened.

The instrument is usually provided with a directing magnet, by which its sensitiveness may be varied to a great extent. This magnet is in the form of a bar, slightly curved, and is of considerable power. It is placed upon a vertical rod passing through its centre, which is fixed above the coil immediately over the needle, in such a manner that it can be turned horizontally so as to follow the movements of the needle, or be removed nearer to or further from it vertically. If it is placed with its south pole over the north pole of the needle, it will add its directive force to that of the earth, and by holding the needle more powerfully in its position, will lessen its sensitiveness. The nearer the magnet approaches the needle the greater will be its power over it, and it can be arranged so as to hold the needle in any desired position. If it is placed in a reverse direction, so as to repel the needle instead of attracting it, it will lessen the attractive force of the earth so as to increase its sensitiveness, and in a certain position will render the galvanometer astatic. When the magnet is too near the needle it repels to the full extent of the scale. If it is raised upon the supporting rod the repelling effect will decrease, until, at a certain distance from the magnet, the spot of light on the scale can be held at zero. The greatest sensibility is obtained at the point at which the slightest lowering of the magnet upon the rod will again repel the needle to the full extent of its swing.

An improvement in this instrument, made by Mr. C. F. Varley, consists in giving the mirror a concave form, silvered upon the back, and thus dispensing with the use of the lens above described.

174. Mode of Working the Atlantic Cables.—Very little has been made public in regard to the precise method employed in signaling through the Atlantic cables. As before remarked, the reflecting galvanometer is employed as a receiving instrument, and by employing deflections on one side of zero to represent dashes, and those on the other side dots, the Morse alphabet is found to answer the purpose admirably. It is said that the two cables have been looped in a metallic circuit without ground connection, and that they have also been worked separately with and without condensers. The latter method is made use of in order to avoid the disturbances generated by what are known as "earth currents."

Different parts of the earth and sea are found to be at different electric potentials. One part is electropositive or electro-negative to another. That is to say, there is the same difference between the two parts of the earth that exists between the two poles of a battery. If, therefore, these two points are joined by a wire, a current will flow through that wire as if from a battery, and this current is termed an earth current, to distinguish it from the current generated by an ordinary voltaic battery. This difference of potential between two given points, such as Newfoundland and Valencia, is not constant but continually varies, causing a corresponding variation in the current it produces. This current and its fluctuations interfere with the signaling current, disturbing the distinctness of the signals. When very rapid changes take place in the electric condition of the earth, it is known as a magnetic storm, and this occasionally interferes with the working of all telegraph lines.

By the method of working with condensers the disturbances from this cause are avoided. The condenser is constructed of alternate layers of tin foil and thin plates of mica, gutta-percha or paper, saturated with paraffine, arranged like the leaves of an interleaved book. Each *alternate* metal plate is connected so as to form two distinct series, insulated from each other, one of which is connected with the line and the other with the earth. By an inductive action, similar to that of the well known Leyden jar, a quantity of electricity, in proportion to the amount of surface exposed, may be accummulated or stored up upon the metallic plates. If, therefore, one series of plates be charged with positive electricity the other series will become negative by induction, and by means of this induction a much larger quantity of electricity may be accumulated than would otherwise be the case.

The manner in which the condenser is made use of in working a cable is as follows:

The sending apparatus consists of a battery, B (Fig. 65), which is permanently connected with the cable through the back contact of a Morse key, K, and the cable is therefore kept constantly charged from this battery. When the key is depressed the cable is placed in connection with the earth at E. The receiving apparatus consists of the reflecting galvanometer G (163), one terminal of which is attached to the cable and the other to one series of plates in the condenser C—the other series being connected with the earth, as shown in the figure. R is a very high resistance, inserted in a wire leading from the point O, between the cable and the galvanometer, so as to allow a very slight but constant leakage from the cable to the earth. The cable is, therefore, charged to the tension of the battery B, and the condenser to a tension equal to that of the point O—but owing to the high resistance at R the tensions are nearly the same. Upon charging the cable with the battery at K a charge of electricity enters the cable, and a quantity sufficient to charge the condenser passes through the galvanometer, deflecting the mirror until the condenser is charged equal to the tension of the point O—when the mirror will return to zero. By putting the cable to earth at K a portion of the charge will be withdrawn, and the tension of the point O lowered below that of the condenser. A portion of the charge of the latter, therefore, flows into the cable, deflecting the galvanometer in the opposite direction. The right and left hand deflections necessary for signaling are therefore produced without reversing the currents, or rendering it necessary to entirely discharge the cable after each signal. This mode of signaling possesses many important advantages over the old method, in point of rapidity of action and freedom from interference by earth currents. The rate of working through the cable by expert operators

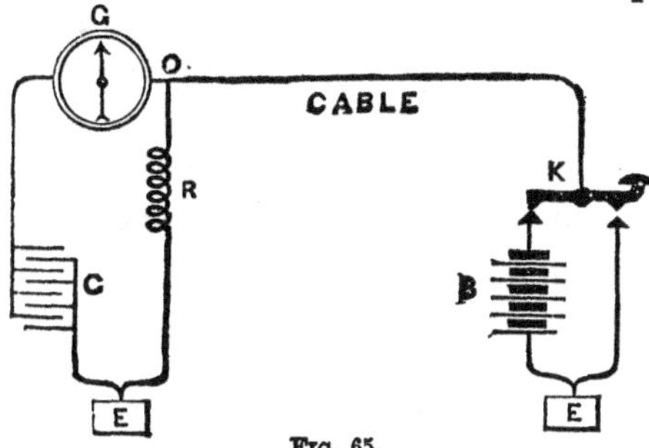

FIG. 65.

is said to average from fifteen to twenty words per minute.

175. VELOCITY OF ELECTRIC SIGNALS.—For many years the velocity of electric signals in passing through a conductor was supposed to be infinitely great, or at least so great as to be incapable of measurement. In 1849, Professor Sears C. Walker, of the United States Coast Survey service, while engaged in measuring longitude by means of the electric telegraph, discovered a perceptible retardation. Experiments between Washington and St. Louis indicated a velocity not far from 16,000 miles per second. Some of the measurements were as low as 11,000 miles per second. On the evening of the 28th of February, 1868, a number of experiments were made by the officers of the Coast Survey, for the purpose of determining accurately the difference in longitude between Cambridge, Mass., and San Francisco, Cal. A wire was connected from Cambridge to San Francisco and back, embracing thirteen repeaters—the whole distance thus traversed by the signals being about 7,000 miles.

The following table shows the time, in hundredths of a second, occupied by a signal in passing from Cambridge to each of the repeating stations and back. The number of repeaters in circuit is also given:

TIME OF TRANSMISSION FROM CAMBRIDGE.

		Seconds.		
To Buffalo and Return		0.10	1	Repeater.
" Chicago	"	0.20	3	"
" Omaha	"	0.33	5	"
" Salt Lake	"	0.54	9	"
" Virginia City	"	0.70	11	"
" San Francisco	"	0.74	13	"

The actual time of transmission from Cambridge to San Francisco and back was estimated not to exceed three tenths of a second, the "armature times" of the thirteen repeaters probably amounting to four or five tenths of a second.

In submarine cables the velocity of signals is considerably less than upon air lines. Prof. Gould, in his experiments upon the Atlantic Cable, found it to be between 7,000 and 8,000 miles per second—being greater when the circuit was composed of the two cables, and less when the earth formed a part of the circuit. His experiments seemed to show that, instead of travelling around the entire circuit in one direction, the electric wave, or polar influence, travelled both ways from the battery, and the signal was received when the two influences met. Experiments made on air lines indicate that an instrument placed at the central point of resistance between the two poles of the battery will record the signal sooner than when placed in any other part of the circuit, it being understood that the two terminal batteries of a telegraph line are in effect but one, being connected by the earth, which is a conductor of infinitely small resistance.

176. SPEED OF TRANSMISSION.—The average rate of transmission, by the most skilful operators upon the Morse apparatus, is about 1,800 words per hour. This has been considerably exceeded, however, by many operators within the past two or three years. On the evening of January 28th, 1868, 2,520 words of Press news were sent from New York to Philadelphia in one hour, and legibly copied by the receiving operator, without a stop or break—the average rate being forty-two, and the maximum rate forty-six words per minute. On the 7th of February following 2,630 words of Press news were sent from Milwaukee, Wis., to St. Paul, Minn., in one hour, the distance being about 400 miles. On the 19th of the same month 1,352 words of Press news were sent from New York to Philadelphia in thirty minutes, the average rate being over forty-five words per minute.

This is believed to be the quickest time on record which has been made in the transmission of regular business by the Morse system. The receiving operator, in all the above cases, copied entirely from the sound of the instrument.

The speed of the printing instrument exceeds that

of the Morse under favorable circumstances. On the 24th of September, 1867, the Combination instrument transmitted from Albany to New York 1,453 words of Press news in thirty-three minutes. It is claimed that, on some occasions, as many as 2,900 words per hour have been transmitted by the House instrument.

177. Comparison of Wire Gauges.—The different sizes of wire employed for telegraphic and other purposes are designated by a series of arbitrary numbers. The system known as the Birmingham gauge is the one in most general use at the present time, but is objectionable, both on account of the irregularity of its gradations and the absence of any authorized standard—wire of the same number from different makers often varying considerably in its size. The American gauge is formed upon a geometrical progression, and it is to be hoped will eventually supersede the old gauge: it is already employed to a considerable extent.* The following table gives the diameter, in thousandths of an inch, of each number in the American and Birmingham gauges:

TABLE OF DIAMETERS OF WIRES.

Number.	American Gauge.	Birmingham Gauge.	Number.	American Gauge.	Birmingham Gauge.
0000	.460	.454	19	.03589	.042
000	.40964	.425	20	.03196	.035
00	.36480	.380	21	.02846	.032
0	.32495	.340	22	.02535	.028
1	.28930	.300	23	.02257	.025
2	.25763	.284	24	.0201	.022
3	.22942	.259	25	.0179	.020
4	.20431	.238	26	.01594	.018
5	.18194	.220	27	.01419	.016
6	.16202	.203	28	.01264	.014
7	.14428	.180	29	.01126	.013
8	.12849	.165	30	.01002	.012
9	.11443	.148	31	.00893	.010
10	.10189	.134	32	.00795	.009
11	.09074	.120	33	.00708	.008
12	.08081	.109	34	.0063	.007
13	.07196	.095	35	.00561	.005
14	.06408	.083	36	.005	.004
15	.05707	.072	37	.00445	....
16	.05082	.065	38	.00396	....
17	.04526	.058	39	.00353	....
18	.0403	.049	40	.00314	....

* This gauge is manufactured by Darling, Brown & Sharpe, of Providence, R. I.

178. Useful Formulæ for Weight and Resistance of Wires.—The following formulæ, from Clark's tables. will be found convenient in telegraphic work:

The *weight* of any iron wire, per statute mile of 5280 feet, is $\frac{d^2}{72.15}$ lbs.; d^2 denoting the square of the diameter of the wire in "mils" or thousandths of an inch.

The *conductivity* of ordinary galvanized iron wire, compared with pure copper 100, averages about 14, or about one seventh that of pure copper.

The *resistance* per statute mile of a galvanized iron wire is about $\frac{395000}{d^2}$ ohms at 60° Fahr.

The resistance of iron wire increases about .35 per cent. for each degree, Fahr.

The *weight* per statute mile of 5280 feet, of any copper wire, is $\frac{d^2}{63.13}$ lbs. A mile of No. 16 wire weighs in practice from 63 to 66 lbs.

The resistance per statute mile of any *pure* copper wire is $\frac{54892}{d^2}$ ohms at 60° Fahr. No. 16 copper wire of good quality has a resistance of about 19 ohms.

The resistance of any pure copper wire l inches in length, weighing n grains, $= \frac{.001516 \times l^2}{n}$ ohms.

The resistance of copper increases as the temperature rises, .21 per cent. for each degree, Fahr.

The conductivity of any copper wire is obtained by multiplying its calculated resistance by 100, and dividing the product by its actual resistance. Pure copper is taken as 100.

179. Conducting Powers of Materials.—According to the experiments of Mr. M. G. Farmer, made some years since, the relative electrical resistance of different metals and fluids at ordinary temperatures is as follows, pure copper being taken as 100:

Copper Wire	1.00	Tin wire	6.80
Silver "	.98	Zinc "	3.70
Gold "	1.13	Brass "	3.88
Iron "	5.63	German Silver Wire	11.30
Lead "	10.76	Nickel "	7.70
Mercury "	50.00	Cadmium "	2.61
Palladium Wire	5.50	Aluminum "	1.75
Platinum "	6.78		

His experiments with fluids gave the following results :

Pure Rain Water	40,653,723,00
Water, 12 parts; Sulphuric Acid, 1 part	1,305,467,00
Sulphate Copper, 1 pound per gallon	18,450,000,00
Saturated solution of common salt	3,173.000,00
" " of sulphate of zinc	17,330,000,00
Nitric Acid, 30 B	1,606,000,00

The following table gives the specific resistance in ohms of various metals and alloys, at 32° Fahr., according to the most recent determinations of Dr. Matthiessen :

NAME OF METALS.	Resistance of wire 1 foot long, weighing 1 grain.	Resistance of wire 1 foot long, 1-1000th inch in diameter.	Approximate per cent. variation in resistance per degree temperature at 20 degrees.
Silver annealed	0.2214	9.936	0.377
" hard drawn	0.2421	9.151	....
Copper annealed	0.2064	9.718	0.388
" hard drawn	0 2106	9.940	....
Gold annealed	0.5849	12.52	0.365
" hard drawn	0.5950	12.74	....
Aluminum annealed	0.06822	17.72	....
Zinc pressed	0.5710	32.22	0.365
Platinum annealed	3.536	55 09	....
Iron annealed	1.2425	59.10	....
Nickel annealed	1.0785	75.78	....
Tin pressed	1.317	80.36	0.365
Lead pressed	3.236	119.39	0.387
Mercury liquid	18.746	600.00	0.072
Platinum silver alloy, hard or annealed, used for standard resistance coils	4.243	148.35	0.031
German silver, hard or annealed, commonly used for resistance coils	2.652	127.32	0.044
Gold silver alloy, 2 parts gold, 1 part silver, hard or annealed	2.391	66.10	0.065

The use of this table is as follows: Suppose it is required to find the resistance at 32° Fahr. of a conductor of pure hard copper, weighing 400 lbs. per knot. This is equivalent to 460 grains per foot. The resistance of a wire weighing one grain is found by the table to be 0.2106, therefore the resistance of a foot of wire weighing 460 grains will be $\frac{0.2106}{460}$, but the resistance of one knot will be 6087 times that of one foot, therefore the resistance required will be $\frac{6087 \times 0.2106}{460} = 2.79$ ohms. If the diameter of the wire be given instead of its weight per knot, the constant is taken from the second column. Thus the resistance at 32° Fahr. of a knot of pure hard drawn copper wire 0.1 inch in diameter would be $\frac{6087 \times 9.94}{10000} = 6.05$. The resistance of wires is materially altered by annealing them, and a rise in temperature increases the resistance of all metals. Dr. Matthiessen found that for all pure metals the increase of resistance between 32° and 212° Fahr. is sensibly the same. The resistance of alloys is much greater than the mean of the metals composing them. They are very useful in the construction of resistance coils.

The highest value which has probably been found for the conducting power of pure copper is sixty times that of pure mercury, according to Sabine. Commercial copper may be considered of good quality when its conducting power is over fifty. Different samples of copper vary greatly in their specific conductivity, as may be seen by the following table, which gives the result of careful determinations by Dr. Matthiessen, the conducting power of pure copper at 59.9° Fahr. being taken as 100.

Lake Superior, native, not fused	98.8 at 59.9°
" " fused (commercial)	92.6 at 59.0°
Burra Burra	88.7 at 57.2°
Best selected	81.3 at 57.5°
Bright copper wire	72.2 at 60.2°
Tough copper	71.0 at 63.1°
Demidoff	59.3 at 54.8°
Rio Tinto	14.2 at 58.6°

Thus Rio Tinto copper possesses no better conducting power than iron. This shows the great importance of testing the conductivity of the wire used in the manufacture of electro-magnets, cables, etc.

180. Internal Resistance of Batteries.—This may be measured by the sine or tangent galvanometer. Place the battery to be measured in circuit with a sine galvanometer giving a certain deflection. Insert resistance till the sine of the deflection becomes half what

it originally was. The total resistance of the circuit is now doubled, and the resistance added is, therefore, equal to the original resistance. Deduct the resistance of the galvanometer and connections from the resistance added, and the remainder is the resistance of the battery.

*Second Method.**—Let D = the deflection obtained with the battery in circuit with a galvanometer whose deflections are proportional, and some resistance r; and d the deflection with some larger resistance R (the resistance of the galvanometer being included in R and r), and let x = the resistance of the battery.

$$\text{Then } D : d :: R + x : r + x$$

$$\text{and } x = \frac{(d \times r) - (D \times r)}{D - d}$$

In using this method any other resistance y may be included with x, and the formulæ becomes—

$$x + y = \frac{(d \times R) - (D \times r)}{D - d,}$$

and by deducting x we get the value of y, or if y be large in comparison with x, the latter may be neglected. By this method one resistance r may be compared with another.

The approximate resistance of the batteries in common use is as follows, according to Mr. Farmer:

Grove	0.41 ohms
Carbon	0.63 "
Daniell	1.70 "

181. Electro-motive Force of Different Batteries.—The following table gives approximately the electro-motive force of various batteries, being the mean of numerous observations taken on a sine galvanometer by Mr. Latimer Clark.† The electro-motive force of batteries is within certain limits very variable, depending on a variety of undetermined causes. It is not much affected by temperature.

* Clark, *Electrical Measurement*, p. 100. † *Electrical Measurement*, p. 103.

Grove's	100
Carbon with bi-chromate solution	107
Daniell's	56
Smee's (when not in action)	57
" (when in action) about	25
Copper and zinc in acid (Wollaston)	46
Sulphate mercury and graphite (Mariè Davy)	76
Chloride silver	62
Chloride lead	30

When connected on short circuit, the electro-motive force of several of the batteries, especially Smee's and Wollaston's, will fall off 50 per cent. or more, owing to the formation of hydrogen on the negative plate. Grove's and Daniell's do not so fall off, because the hydrogen is reduced by the nitric acid in one case and by the oxygen in the other.

182. MEASUREMENT OF ELECTRO-MOTIVE FORCE.*—When a number of cells are joined up in circuit with, but in opposition to, a number of other cells with a galvanometer inserted, by adjusting the number of cells so that no current passes, the relative electro-motive force of the two batteries may be determined.

Second Method.—Call the electro-motive forces of the two batteries E and E′; join them up successively in circuit with the same galvanometer, and by varying the resistance, cause them both to give the same deflection; their forces will then be in direct proportion to the *total* resistances in circuit in each case, or

$$E' = E \times \frac{R'}{R}$$

where R represents the resistance with E (including that of battery, galvanometer, and the adjustable resistance) and R′ with E′.

183. FORCES OF ELECTRO-MAGNETS.—The laws which govern the forces of electro-magnets have been investigated by Lenz, Jacobi and Müller.

Let M = the magnetic force of the electro-magnet.
n = the number of convolutions of wire.
d = the diameter of the soft iron core.
Q = the quantity of electricity in circulation.
and c a constant multiplier.

Then $M = c\, n\, Q \sqrt{d}$.

* Clark, *Electrical Measurement*, p. 103.

This law only holds good for bars of iron whose length is considerably greater than their diameter, for feeble currents of electricity, and under the supposition that the number of convolutions of wire is not so great as materially to diminish the influence exerted by the outer coils upon the bar of iron. These conditions are fulfilled in the electro-magnets used for telegraphic purposes.

It will be noticed, in the above formulæ, that M increases directly as Q and as n, but Q decreases as n increases, supposing the electric force to remain constant. Hence it is evident that a certain proportion between the resistance of the wire and that of the remaining portions of the circuit must be preserved to obtain the maximum magnetic force. This relation is found to be the following:

*When the resistance of the coils of the electro-magnet is equal to the resistance of the rest of the circuit, i. e., the conducting wire and battery, the magnetic force is a maximum.**

The application of this law to a telegraphic circuit would be to make the sum of the resistances of all the magnet coils in circuit equal to the resistance of the line and batteries, but as in practice the resistance of a telegraphic circuit varies, being considerably reduced by defective insulation, the total resistance of the instruments should be less than that of the line when in good condition, to attain the best results during unfavorable weather.

ELECTRICAL FORMULÆ.

184. OHM'S LAW.—Let C = the quantity, or strength, or force, or intensity of the current, as it is variously called.

Let n = the number of cells.
" E = the electro-motive force in each cell.
" R = the internal resistance of each cell.
" r = the resistances exterior to the battery.

Then
$$C = \frac{n E}{n R + r}.$$

* Noad's Students' Text-book of Electricity, p. 277.

185. PARALLEL OR DERIVED CIRCUITS.—1. The joint resistance of any two parallel or derived circuits, whose resistances = a and b, is equal to their product divided by their sum, or

$$R = \frac{a\,b}{a + b}$$

2. The joint resistance of any three circuits, a, b and c, is

$$R = \frac{a\,b\,c}{a\,b + b\,c + a\,c}$$

3. The joint resistance of any number of circuits is obtained by adding their reciprocals together, thus:

$$R = \frac{1}{\frac{1}{a} + \frac{1}{b} + \frac{1}{c}}$$

186. GALVANOMETERS AND SHUNTS.—1. The joint resistance of a galvanometer and shunt is as follows:

Let g = resistance of galvanometer.
s = resistance of shunt.

$$\text{Then } R = \frac{g\,s}{g + s}$$

2. The multiplying power of any shunt is equal to

$$\frac{g + s}{s}, \text{ or } \frac{g}{s} + 1$$

3. To prepare a shunt having some definite multiplying power, for example 10 100 or 1,000,

Let n = the multiplying power requred,

$$\text{Then } s = \frac{g}{n - 1}$$

187. FORMULA FOR THE LOOP TEST (127 .—Let x = resistance of shortest part of the loop.

y = resistance of longest part.
L = total resistance of both.
R = resistance added to shortest part, to make it equal to the longer.

$$\text{Then } x + y = L.$$
$$y = x + R.$$
$$\text{and } X = \frac{L - R}{2}$$

188. BLAVIER'S FORMULA FOR LOCATING A FAULT (128).—Let R = resistance of line when in good order, S = resistance of defective line when distant end is to ground, and T the resistance when it is disconnected or open at distant end.

The distance (x) of the fault from the testing station will be

$$x = S - \sqrt{S^2 + TR - TS - RS},$$
$$\text{or } x = S - \sqrt{(R - S) \times (T - S)},$$

and the resistance of the fault (z) will be

$$z = T - S + \sqrt{S^2 + TR - TS - RS}$$
$$\text{or } z = T - S + \sqrt{(R - S) \times (T - S)}$$

189. MEASURES OF RESISTANCE.—1.0456 Siemen's units = 1 ohm. To convert Siemen's units into ohms. multiply by .9564.

1 Varley's unit = 25 ohms.
1 Megohm = 1,000,000 ohms.
1 Microhm = $\frac{1}{1,000,000}$ ohms.

STRAIN OF SUSPENDED WIRES.*—The ordinary dip of line wires, for a span of 80 yards, is about 18 inches in mild weather; this gives with No. 8 wire a strain of 420 lbs., its breaking weight being about 1,300 lbs.—(*Culley.*)

The strain varies directly as the weight of the wire, and inversely as the dip or versine; it increases as the square of the span if the dip be constant; but to preserve a given strain the dip or versine must increase as the square of the span, or,

$$L^2 : l^2 :: V : v.$$

The strain is greater at the point of suspension than at the lowest point of the span, by a quantity (equal to the weight of a length of wire of the same height as the versine) which may be neglected in practice. Calling l the length of the span in feet, w the weight in

---

* Clark. *Resistance Measurement.* p. 154.

cwts. of one statute mile, v the versine in inches, and s the strain in lbs.,

$$\text{Strain} = \frac{l^2 \times w}{31.43 \times v} \text{ lbs. approximately.}$$

$$\text{and dip} = \frac{l^2 \times w}{31.43 \times s} \text{ inches.}$$

When both supports are of the same height the lowest point in the curve will be in the centre of the span; but if one support be higher than the other the lowest point will be near the lower support, so that the greater portion of the weight is borne by the higher pole. In calculating the strain the wire should be considered as if prolonged beyond the lower end to a point equal in height to the upper one, and the strain will be proportional to the length thus increased, or to twice the distance from the top to the bottom of the dip.

The weight of a wire increases with its strength, the quality being the same. The advantage of using thin wire for long spans is only in diminishing the weight upon the supports.

Iron expands $\frac{1}{14616}$ of its length, or about $4\frac{1}{10}$ inches per mile for every ten degrees of heat.—(*Culley.*)

THE END.

INDEX.

SCIENTIFIC BOOKS.

FRANCIS' (J. B.) Hydraulic Experiments. Lowell Hydraulic Experiments—being a Selection from Experiments on Hydraulic Motors, on the Flow of Water over Weirs, and in Open Canals of Uniform Rectangular Section, made at Lowell, Mass. By J. B. FRANCIS, Civil Engineer. Second edition, revised and enlarged, including many New Experiments on Gauging Water in Open Canals, and on the Flow through Submerged Orifices and Diverging Tubes. With 23 copperplates, beautifully engraved, and about 100 new pages of text. 1 vol., 4to. Cloth. $15.

Most of the practical rules given in the books on hydraulics have been determined from experiments made in other countries, with insufficient apparatus, and on such a minute scale, that in applying them to the large operations arising in practice in this country, the engineer cannot but doubt their reliable applicability. The parties controlling the great water-power furnished by the Merrimack River at Lowell, Massachusetts, felt this so keenly, that they have deemed it necessary, at great expense, to determine anew some of the most important rules for gauging the flow of large streams of water, and for this purpose have caused to be made, with great care, several series of experiments on a large scale, a selection from which are minutely detailed in this volume.

The work is divided into two parts—PART I., on hydraulic motors, includes ninety-two experiments on an improved Fourneyron Turbine Water-Wheel, of about two hundred horse-power, with rules and tables for the construction of similar motors:—Thirteen experiments on a model of a centre-vent water-wheel of the most simple design, and thirty-nine experiments on a centre vent water-wheel of about two hundred and thirty horse-power.

PART II. includes seventy-four experiments made for the purpose of determining the form of the formula for computing the flow of water over weirs; nine experiments on the effect of back water on the flow over weirs; eighty-eight experiments made for the purpose of determining the formula for computing the flow over weirs of regular or standard forms, with several tables of comparisons of the new formula with the results obtained by former experimenters; five experiments on the flow over a dam in which the crest was of the same form as that built by the Essex Company across the Merrimack River at Lawrence, Massachusetts; twenty-one experiments on the effect of observing the depths of water on a weir at different distances from the weir; an extensive series of experiments made for the purpose of determining rules for gauging streams of water in open canals, with tables for facilitating the same; and one hundred and one experiments on the discharge of water through submerged orifices and diverging tubes, the whole being fully illustrated by twenty-three double plates engraved on copper.

In 1855 the proprietors of the Locks and Canals on Merrimack River, at whose expense most of the experiments were made, being willing that the public should share the benefits of the scientific operations promoted by them, consented to the publication of the first edition of this work, which contained a selection of the most important hydraulic experiments made at Lowell up to that time. In this second edition the principal hydraulic experiments made there, subsequent to 1855, have been added, including the important series above mentioned, for determining rules for the gauging the flow of water in open canals, and the interesting series on the flow through a submerged Venturi's tube, in which a larger flow was obtained than any we find recorded.

FRANCIS (J. B.) On the Strength of Cast-Iron Pillars, with Tables for the use of Engineers, Architects, and Builders. By JAMES B. FRANCIS, Civil Engineer. 1 vol., 8vo. Cloth. $2.

WILLIAMSON (R. S.) On the Use of the Barometer on Surveys and Reconnaissances. Part I. Meteorology in its Connection with Hypsometry. Part II. Barometric Hypsometry. By R. S. WILLIAMSON, Bvt. Lieut.-Col. U. S. A., Major Corps of Engineers. With Illustrative Tables and Engravings. Paper No. 15, Professional Papers, Corps of Engineers. 1 vol., 4to. Cloth. $15.

"SAN FRANCISCO, CAL., *Feb.* 27, 1867.

"Gen. A. A. HUMPHREYS, Chief of Engineers, U. S. Army:

"GENERAL—I have the honor to submit to you, in the following pages, the results of my investigations in meteorology and hypsometry, made with the view of ascertaining how far the barometer can be used as a reliable instrument for determining altitudes on extended lines of survey and reconnaissances. These investigations have occupied the leisure permitted me from my professional duties during the last ten years, and I hope the results will be deemed of sufficient value to have a place assigned them among the printed professional papers of the United States Corps of Engineers. Very respectfully, your obedient servant,

"R. S. WILLIAMSON,

"Bvt. Lt.-Col. U. S. A., Major Corps of U. S. Engineers."

TUNNER (P.) A Treatise on Roll-Turning for the Manufacture of Iron. By PETER TUNNER. Translated and adapted. By JOHN B. PEARSE, of the Pennsylvania Steel Works. With numerous engravings and wood-cuts. 1 vol., 8vo., with 1 vol. folio of plates. Cloth. $10

SHAFFNER (T. P.) Telegraph Manual. A Complete History and Description of the Semaphoric, Electric, and Magnetic Telegraphs of Europe, Asia, and Africa, with 625 illustrations. By TAL. P. SHAFFNER, of Kentucky. New edition. 1 vol., 8vo. Cloth. 850 pp. $6.50.

MINIFIE (WM.) Mechanical Drawing. A Text-Book of Geometrical Drawing for the use of Mechanics and Schools, in which the Definitions and Rules of Geometry are familiarly explained; the Practical Problems are arranged, from the most simple to the more complex, and in their description technicalities are avoided as much as possible. With illustrations for Drawing Plans, Sections, and Elevations of Buildings and Machinery; an Introduction to Isometrical Drawing, and an Essay on Linear Perspective and Shadows. Illustrated with over 200 diagrams engraved on steel. By WM. MINIFIE, Architect. Seventh edition. With an Appendix on the Theory and Application of Colors. 1 vol., 8vo. Cloth. $4.

It is the best work on Drawing that we have ever seen, and is especially a text-book of Geometrical Drawing for the use of Mechanics and Schools. No young Mechanic, such as a Machinist, Engineer, Cabinet-Maker, Millwright, or Carpenter should be without it."—*Scientific American.*

"One of the most comprehensive works of the kind ever published, and cannot but possess great value to builders. The style is at once elegant and substantial."—*Pennsylvania Inquirer*

"Whatever is said is rendered perfectly intelligible by remarkably well-executed diagrams on steel, leaving nothing for mere vague supposition; and the addition of an introduction to isometrical drawing, linear perspective, and the projection of shadows, winding up with a useful index to technical terms."—*Glasgow Mechanics' Journal.*

☞ The British Government has authorized the use of this book in their schools of art at Somerset House, London, and throughout the kingdom.

MINIFIE (WM.) Geometrical Drawing. Abridged from the octavo edition, for the use of Schools. Illustrated with 48 steel plates. New edition, enlarged. 1 vol., 12mo, cloth. $2.

"It is well adapted as a text-book of drawing to be used in our High Schools and Academies where this useful branch of the fine arts has been hitherto too much neglected."—*Boston Journal.*

PEIRCE'S SYSTEM OF ANALYTIC MECHANICS. Physical and Celestial Mechanics, by BENJAMIN PEIRCE, Perkins Professor of Astronomy and Mathematics in Harvard University, and Consulting Astronomer of the American Ephemeris and Nautical Almanac. Developed in four systems of Analytic Mechanics, Celestial Mechanics, Potential Physics, and Analytic Morphology. 1 vol., 4to. Cloth. $10.

GILLMORE. Practical Treatise on Limes, Hydraulic Cements, and Mortars. Papers on Practical Engineering, U. S. Engineer Department, No. 9, containing Reports of numerous experiments conducted in New York City, during the years 1858 to 1861, inclusive. By Q. A. GILLMORE, Brig.-General U. S. Volunteers, and Major U. S. Corps of Engineers. With numerous illustrations. One volume, octavo. Cloth. $4.

ROGERS (H. D.) Geology of Pennsylvania. A complete Scientific Treatise on the Coal Formations. By HENRY D. ROGERS, Geologist. 3 vols., 4to., plates and maps. Boards. $30.00.

BURGH (N. P.) Modern Marine Engineering, applied to Paddle and Screw Propulsion. Consisting of 36 colored plates, 259 Practical Woodcut Illustrations, and 403 pages of Descriptive Matter, the whole being an exposition of the present practice of the following firms: Messrs. J. Penn & Sons; Messrs. Maudslay, Sons, & Field; Messrs. James Watt & Co.; Messrs. J. & G. Rennie; Messrs. R. Napier & Sons; Messrs. J. & W. Dudgeon; Messrs. Ravenhill & Hodgson; Messrs. Humphreys & Tenant; Mr. J. T. Spencer, and Messrs. Forrester & Co. By N. P. BURGH, Engineer. In one thick vol., 4to. Cloth. $25.00. Half morocco. $30.00.

KING. Lessons and Practical Notes on Steam, the Steam-Engine, Propellers, &c., &c., for Young Marine Engineers, Students, and others. By the late W. R. KING, U. S. N. Revised by Chief-Engineer J. W. KING, U. S. Navy. Twelfth edition, enlarged. 8vo. Cloth. $2.

WARD. Steam for the Million. A Popular Treatise on Steam and its Application to the Useful Arts, especially to Navigation. By J. H. WARD, Commander U. S. Navy. New and revised edition. 1 vol., 8vo. Cloth. $1.

WALKER. Screw Propulsion. Notes on Screw Propulsion, its Rise and History. By Capt. W. H. WALKER, U. S. Navy. 1 vol., 8vo. Cloth. 75 cents.

THE STEAM-ENGINE INDICATOR, and the Improved Manometer Steam and Vacuum Gauges: Their Utility and Application. By PAUL STILLMAN. New edition. 1 vol., 12mo. Flexible cloth. $1.

ISHERWOOD. Engineering Precedents for Steam Machinery. Arranged in the most practical and useful manner for Engineers. By B. F. ISHERWOOD, Civil Engineer U. S. Navy. With illustrations. Two volumes in one. 8vo. Cloth. $2.50.

Scientific Books.

POOK'S METHOD OF COMPARING THE LINES AND DRAUGHTING VESSELS PROPELLED BY SAIL OR STEAM, including a Chapter on Laying off on the Mould-Loft Floor. By SAMUEL M. POOK, Naval Constructor. 1 vol., 8vo. With illustrations. Cloth. $5.

SWEET (S. H.) Special Report on Coal; showing its Distribution, Classification and Cost delivered over different routes to various points in the State of New York, and the principal cities on the Atlantic Coast. By S. H. SWEET. With maps. 1 vol., 8vo. Cloth. $3.

ALEXANDER (J. H.) Universal Dictionary of Weights and Measures, Ancient and Modern, reduced to the standards of the United States of America. By J. H. ALEXANDER. New edition. 1 vol., 8vo. Cloth. $3.50.

"As a standard work of reference this book should be in every library; it is one which we have long wanted, and it will save us much trouble and research."—*Scientific American*

CRAIG (B. F.) Weights and Measures. An Account of the Decimal System, with Tables of Conversion for Commercial and Scientific Uses. By B. F. CRAIG, M. D. 1 vol., square 32mo. Limp cloth. 50 cents.

"The most lucid, accurate, and useful of all the hand-books on this subject that we have yet seen. It gives forty-seven tables of comparison between the English and French denominations of length, area, capacity, weight, and the centigrade and Fahrenheit thermometers, with clear instructions how to use them; and to this practical portion, which helps to make the transition as easy as possible, is prefixed a scientific explanation of the errors in the metric system, and how they may be corrected in the laboratory."—*Nation.*

BAUERMAN. Treatise on the Metallurgy of Iron, containing outlines of the History of Iron manufacture, methods of Assay, and analysis of Iron Ores, processes of manufacture of Iron and Steel, etc., etc. By H. BAUERMAN. First American edition. Revised and enlarged, with an appendix on the Martin Process for making Steel, from the report of Abram S. Hewitt. Illustrated with numerous wood engravings. 12mo. Cloth. $2.50.

"This is an important addition to the stock of technical works published in this country. It embodies the latest facts, discoveries, and processes connected with the manufacture of iron and steel, and should be in the hands of every person interested in the subject, as well as in all technical and scientific libraries."—*Scientific American.*

HARRISON. Mechanic's Tool Book, with practical rules and suggestions, for the use of Machinists, Iron Workers, and others. By W. B. HARRISON, associate editor of the "American Artisan." Illustrated with 44 engravings. 12mo. Cloth. $2.50.

"This work is specially adapted to meet the wants of Machinists and workers in iron generally. It is made up of the work-day experience of an intelligent and ingenious mechanic, who had the faculty of adapting tools to various purposes. The practicability of his plans and suggestions are made apparent even to the unpracticed eye by a series of well-executed wood engravings."—*Philadelphia Inquirer.*

PLYMPTON. The Blow-Pipe: A System of Instruction in its practical use, being a graduated course of Analysis for the use of students, and all those engaged in the Examination of Metallic Combinations. Second edition, with an appendix and a copious index. By GEORGE W. PLYMPTON, of the Polytechnic Institute, Brooklyn. 12mo. Cloth. $2.

"This manual probably has no superior in the English language as a text-book for beginners, or as a guide to the student working without a teacher. To the latter many illustrations of the utensils and apparatus required in using the blow-pipe, as well as the fully illustrated description of the blow-pipe flame, will be especially serviceable."—*New York Teacher.*

NUGENT. Treatise on Optics: or, Light and Sight, theoretically and practically treated; with the application to Fine Art and Industrial Pursuits. By E. NUGENT. With one hundred and three illustrations. 12mo. Cloth. $2.

"This book is of a practical rather than a theoretical kind, and is designed to afford accurate and complete information to all interested in applications of the science."—*Round Table.*

SILVERSMITH (Julius). A Practical Hand-Book for Miners, Metallurgists, and Assayers, comprising the most recent improvements in the disintegration, amalgamation, smelting, and parting of the Precious Ores, with a Comprehensive Digest of the Mining Laws. Greatly augmented, revised, and corrected. By JULIUS SILVERSMITH. Fourth edition. Profusely illustrated. 1 vol., 12mo. Cloth. $3.

LARRABEE'S CIPHER AND SECRET LETTER AND TELEGRAPHIC CODE. By C. S. LARRABEE. 18mo. Cloth. $1.

BRÜNNOW. Spherical Astronomy. By F. BRÜNNOW, Ph. Dr. Translated by the Author from the Second German edition. 1 vol., 8vo. Cloth. $6.50.

CHAUVENET (Prof. Wm.) New method of Correcting Lunar Distances, and Improved Method of Finding the Error and Rate of a Chronometer, by equal altitudes. By WM. CHAUVENET, LL.D. 1 vol., 8vo. Cloth. $2.

POPE. Modern Practice of the Electric Telegraph. A Handbook for Electricians and Operators. By FRANK L. POPE. Fourth edition. Revised and enlarged, and fully illustrated. 8vo. Cloth. $2.

GAS WORKS OF LONDON. By ZERAH COLBURN. 12mo. Boards. 60 cents.

HEWSON. Principles and Practice of Embanking Lands from River Floods, as applied to the Levees of the Mississippi. By WILLIAM HEWSON, Civil Engineer. 1 vol., 8vo. Cloth. $2.

"This is a valuable treatise on the principles and practice of embanking lands from river floods, as applied to Levees of the Mississippi, by a highly intelligent and experienced engineer. The author says it is a first attempt to reduce to order and to rule the design, execution, and measurement of the Levees of the Mississippi. It is a most useful and needed contribution to scientific literature."—*Philadelphia Evening Journal.*

WEISBACH'S MECHANICS. New and revised edition. A Manual of the Mechanics of Engineering, and of the Construction of Machines. By JULIUS WEISBACH, PH. D. Translated from the fourth augmented and improved German edition, by ECKLEY B. COXE, A. M., Mining Engineer. Vol. I.—Theoretical Mechanics. 1 vol. 8vo, 1,100 pages, and 902 wood-cut illustrations, printed from electrotype copies of those of the best German edition. $10.

ABSTRACT OF CONTENTS.—Introduction to the Calculus—The General Principles of Mechanics—Phoronomics, or the Purely Mathematical Theory of Motion—Mechanics, or the General Physical Theory of Motion—Statics of Rigid Bodies—The Application of Statics to Elasticity and Strength—Dynamics of Rigid Bodies—Statics of Fluids—Dynamics of Fluids—The Theory of Oscillation, etc.

"The present edition is an entirely new work, greatly extended and very much improved. It forms a text-book which must find its way into the hands, not only of every student, but of every engineer who desires to refresh his memory or acquire clear ideas on doubtful points."—*The Technologist.*

HUNT (R. M.) Designs for the Gateways of the Southern Entrances to the Central Park. By RICHARD M. HUNT. With a description of the designs. 1 vol., 4to. Illustrated. Cloth. $5.

SILVER DISTRICTS OF NEVADA. 8vo., with map. Paper. 35 cents.

McCORMICK (R. C.) Arizona: Its Resources and Prospects. By Hon. R. C. McCORMICK. With map. 8vo. Paper. 25 cents.

SIMM'S LEVELLING. A Treatise on the Principles and Practice of Levelling, showing its application to purposes of Railway Engineering and the Construction of Roads, &c. By FREDERICK W. SIMMS, C. E. From the fifth London edition, revised and corrected, with the addition of Mr. Law's Practical Examples for Setting Out Railway Curves. Illustrated with three lithographic plates and numerous wood-cuts. 8vo. Cloth. $2.50.

SAELTZER. Treatise on Acoustics in Connection with Ventilation. With a new theory, based on an important discovery, of facilitating clear and intelligible sound in any building. By Alexander Saeltzer. 12mo. In press.

BURT. Key to the Solar Compass, and Surveyor's Companion; comprising all the Rules necessary for use in the field. By W. A. BURT, U. S. Deputy Surveyor. Second edition. Pocket-book form, tuck, $2.50.

GILLMORE. Coignet Béton and other Artificial Stone. By Q. A. GILLMORE. 9 Plates, Views, &c. 8vo, cloth, $2.50.

AUCHINCLOSS. Application of the Slide Valve and Link Motion to Stationary, Portable, Locomotive, and Marine Engines, with new and simple methods for proportioning the parts. By WILLIAM S AUCHINCLOSS, Civil and Mechanical Engineer. Designed as a handbook for Mechanical Engineers, Master Mechanics, Draughtsmen, and Students of Steam Engineering. All dimensions of the valve are found with the greatest ease by means of a PRINTED SCALE, and proportions of the link determined *without* the assistance of a model. Illustrated by 37 woodcuts and 21 lithographic plates, together with a copperplate engraving of the Travel Scale. 1 vol. 8vo. Cloth. $3.

HUMBER'S STRAINS IN GIRDERS. A Handy Book for the Calculation of Strains in Girders and Similar Structures, and their Strength, consisting of Formulæ and Corresponding Diagrams, with numerous details for practical application. By WILLIAM HUMBER. 1 vol. 18mo. Fully illustrated. Cloth. $2.50.

GLYNN ON THE POWER OF WATER, as applied to drive Flour Mills, and to give motion to Turbines and other Hydrostatic Engines. By JOSEPH GLYNN, F. R. S. Third edition, revised and enlarged, with numerous illustrations. 12mo. Cloth. $1.25.

THE KANSAS CITY BRIDGE, with an Account of the Regimen of the Missouri River, and a description of the Methods used for Founding in that River. By O. CHANUTE, Chief Engineer, and GEORGE MORISON, Assistant Engineer. Illustrated with five lithographic views and 12 plates of plans. 4to. Cloth. $6.

TREATISE ON ORE DEPOSITS. By BERNHARD VON COTTA, Professor of Geology in the Royal School of Mines, Freidberg, Saxony. Translated from the second German edition, by FREDERICK PRIME, Jr., Mining Engineer, and revised by the author, with numerous illustrations. 1 vol. 8vo. Cloth, $4.

A TREATISE ON THE RICHARDS STEAM-ENGINE INDICATOR, with directions for its use. By CHARLES T. PORTER. Revised, with notes and large additions as developed by American Practice, with an Appendix containing useful formulæ and rules for Engineers. By F. W. BACON, M. E., member of the American Society of Civil Engineers. 12mo. Illustrated. Cloth. $1

THE ART OF GRAINING. How Acquired and How Produced. By Charles Pickert and Abraham Metcalf. 8vo. Beautifully Illustrated. Tinted paper. In press.

INVESTIGATIONS OF FORMULAS, for the Strength of the Iron Parts of Steam Machinery. By J. D. VAN BUREN, Jr., C. E. 1 vol. 8vo. Illustrated. Cloth. $2.

SUBMARINE BLASTING in Boston Harbor, Massachusetts—Removal of Tower and Corwin Rocks. By JOHN G. FOSTER, U. S. A. Illustrated with 7 plates. 4to. Cloth. $3.50.

KIRKWOOD. Report on the Filtration of River Waters, for the supply of Cities, as practised in Europe, made to the Board of Water Commissioners of the City of St. Louis. By JAMES P. KIRKWOOD. Illustrated by 30 double-plate engravings. 4to. Cloth. $15.00.

LECTURE NOTES ON PHYSICS. By ALFRED M. MAYER, Ph. D., Professor of Physics in the Lehigh University. 1 vol. 8vo. Cloth. $2.

THE PLANE TABLE, and its Uses in Topographical Surveying. From the Papers of the U. S. Coast Survey. 8vo. Cloth. $2.

DIEDRICHS. Theory of Strains; a Compendium for the Calculation and Construction of Bridges, Roofs, and Cranes, with the Application of Trigonometrical Notes. By John Diedrichs. Illustrated by numerous Plates and Diagrams. 8vo, cloth. $5.00.

WILLIAMSON. Practical Tables in Meteorology and Hypsometry, in connection with the use of the Barometer. By Col. R. S Williamson, U. S. A. 1 vol. 4to, flexible cloth. $2.50.

CULLEY. A Hand-Book of Practical Telegraphy. By R. S. CULLEY. Engineer to the Electric and International Telegraph Company, Fourth edition, revised and enlarged. 8vo. Illustrated. Cloth. $5.

RANDALL'S QUARTZ OPERATOR'S HAND-BOOK. By P. M. RANDALL. New edition, revised and enlarged. Fully illustrated. 12mo. Cloth. $2.00.

GOUGE. New System of Ventilation, which has been thoroughly tested under the patronage of many distinguished persons. By HENRY A. GOUGE. Third edition, enlarged. With many illustrations. 8vo. Cloth. $2.

PLATTNER'S BLOW-PIPE ANALYSIS. A Complete Guide to Qualitative and Quantitative Examinations with the Blow-Pipe. Revised and enlarged by Prof. RICHTER, Freiberg. Translated from the latest German edition by HENRY B. CORNWALL, A. M., E. M. 8vo.

GRÜNER. The Manufacture of Steel. By M. L. Grüner. Translated from the French by Lenox Smith, A.M., E.M. With an Appendix on the Bessemer Process in the United States, by the Translator. Illustrated. 8vo. In press

www.ingramcontent.com/pod-product-compliance
Lightning Source LLC
LaVergne TN
LVHW011225110826
845150LV00006B/1555